Advances in Experimental Medicine and Biology

Neuroscience and Respiration

Volume 1020

More information about this series at http://www.springer.com/series/13457

Mieczyslaw Pokorski
Editor

Clinical Research and Practice

Springer

Editor
Mieczyslaw Pokorski
Opole Medical School
Opole, Poland

ISSN 0065-2598 ISSN 2214-8019 (electronic)
Advances in Experimental Medicine and Biology
ISBN 978-3-319-65444-7 ISBN 978-3-319-65445-4 (eBook)
DOI 10.1007/978-3-319-65445-4

Library of Congress Control Number: 2017949360

Printed on acid-free paper

This Springer imprint is published by Springer Nature
The registered company is Springer International Publishing AG
The registered company address is: Gewerbestrasse 11, 6330 Cham, Switzerland

Preface

The book series Neuroscience and Respiration presents contributions by expert researchers and clinicians in the multidisciplinary areas of medical research and clinical practice. Particular attention is focused on pulmonary disorders as the respiratory tract is upfront at the first line of defense for organisms against pathogens and environmental or other sources of toxic or disease-causing effects. The articles provide timely overviews of contentious issues or recent advances in the diagnosis, classification, and treatment of the entire range of diseases and disorders, both acute and chronic. The texts are thought as a merger of basic and clinical research dealing with biomedicine at both the molecular and functional levels and with the interactive relationship between respiration and other neurobiological systems, such as cardiovascular function, immunogenicity, endocrinology and humoral regulation, and the mind-to-body connection. The authors focus on modern diagnostic techniques and leading-edge therapeutic concepts, methodologies, and innovative treatments. The action and pharmacology of existing drugs and the development and evaluation of new agents are the heady area of research. Practical, data-driven options to manage patients are considered. New research is presented regarding older drugs, performed from a modern perspective or from a different pharmacotherapeutic angle. The introduction of new drugs and treatment approaches in both adults and children is also discussed.

Body functions, including lung ventilation and its regulation, are ultimately driven by the brain. However, neuropsychological aspects of disorders are still mostly a matter of conjecture. After decades of misunderstanding and neglect, emotions have been rediscovered as a powerful modifier or even the probable cause of various somatic disorders. Today, the link between stress and health is undeniable. Scientists accept a powerful psychological connection that can directly affect our quality of life and health span. Psychological approaches, which can decrease stress, can play a major role in disease therapy.

Neuromolecular and carcinogenetic aspects relating to gene polymorphism and epigenesis, involving both heritable changes in the nucleotide sequence and functionally relevant changes to the genome that do not involve a change in the nucleotide sequence, leading to disorders, are also tackled.

Clinical advances stemming from molecular and biochemical research are but possible if research findings are translated into diagnostic tools, therapeutic procedures, and education, effectively reaching physicians and patients. All this cannot be achieved without a multidisciplinary, collaborative, bench-to-bedside approach involving both researchers and clinicians. The role of science in shaping medical knowledge and transforming it into practical care is undeniable.

Concerning respiratory disorders, their societal and economic burden has been on the rise worldwide, leading to disabilities and shortening of life-span. COPD alone causes more than three million deaths globally each year. Concerted efforts are required to improve this situation, and part of those efforts are gaining insights into the underlying mechanisms of disease and staying abreast with the latest developments in diagnosis and treatment regimens. It is hoped that the articles published in this series will assume a leading position as a source of information on interdisciplinary medical research advancements, addressing the needs of medical professionals and allied health care workers, and become a source of reference and inspiration for future research ideas.

I would like to express my deep gratitude to Paul Roos, Tanja Koppejan, and Cynthia Kroonen of Springer SBM NL for their genuine interest in making this scientific endeavor come through and in the expert management of the production of this novel book series.

Mieczyslaw Pokorski

Contents

Advs Exp. Medicine, Biology - Neuroscience and Respiration (2017) 32: 1–6
DOI 10.1007/5584_2017_4

Published online: 3 March 2017

Regional Diversification of Influenza Activity in Poland During the 2015/16 Epidemic Season

K. Szymański, D. Kowalczyk, K. Cieślak, and L.B. Brydak

Abstract

The National Influenza Center (NIC) at the Department of Influenza Research of the National Institute for Public Health–National Institute of Hygiene (NIPH-NIH) participates in the Global Influenza Surveillance and Response System (GISRS) and continuously coordinates epidemiological and virological surveillance of influenza in Poland. The aim of this study was to determine the regional differences of influenza activity in Poland in the 2015/16 epidemic season. The influenza surveillance involved 16 administrative districts in which there are Voivodeship (province) Sanitary Epidemiological Stations set up to report influenza and influenza-like illness among the Polish population. Over 8000 specimens were tested in the season with regard to the respiratory viral infections in all regions investigated. The circulation of influenza viruses A and B was confirmed, with the subtype A/H1N1/pdm09 being predominant in the Pomerania, Podlaskie, Subcarpathian, Lubuskie, Silesian, and Warmian-Masuria provinces. The influenza-like virus occurred in individual cases, except for respiratory syncytial virus that also was detected in the Greater Poland and Warmia-Masuria provinces. The highest incidence of cases and suspected cases of influenza was recorded in Pomerania and the lowest one in Lubuskie provinces. The knowledge of regional differences in influenza activity is important for streamlining the distribution of preventive, therapeutic, and economic resources to combat the epidemic.

K. Szymański (✉), D. Kowalczyk, K. Cieślak, and L.B. Brydak
Department of Influenza Research, National Influenza Center, National Institute of Public Health–National Institute of Hygiene, 24 Chocimska Street 24, 00-791 Warsaw, Poland
e-mail: kszymanski@pzh.gov.pl

Keywords
Epidemic • Influenza infection • Regional distribution • Respiratory infection • Respiratory viruses • Vaccine

1 Introduction

Influenza viruses are the most common cause of respiratory infections, resulting in high morbidity and mortality (Taubenberger and Morens 2008). Influenza is an infectious disease caused by influenza virus type A or B. The occurrence of influenza, whose viruses are characterized by great seasonal antigenic changeability, may vary in different districts of a country. In Poland, epidemiological surveillance of influenza was implemented in collaboration with primary care physicians and laboratories of 16 Voivodeship (province) Sanitary Epidemiological Station (VSES) as of the 2004/05 epidemic season. The information on influenza and influenza-like infections has been gathered in all districts by means of both Sentinel and non-Sentinel systems. Since the 2013/14 epidemic season, data consisting of the number of samples tested, suspected and confirmed cases of infection, hospitalizations, and deaths in each province are stratified into seven age-groups (Hallmann-Szelińska et al. 2016a, b; Bednarska et al. 2015). These data are then further elaborated and randomly checked for correctness of viral identification in the National Influenza Center (NIC) at the National Institute of Public Health-National Institute of Health (NIPH-NIH 2016). The NIC serves as a reference center for individual virological VSES laboratories of the country.

The aim of the present study was to determine the regional differences of influenza activity in Poland in the 2015/16 epidemic season. The savvy of such differences might be of help in setting apart resources for preventive and therapeutic actions to combat the epidemic.

2 Methods

2.1 Clinical Specimens

The study was approved by an institutional Ethics Committee and it was conducted in accordance with the Declaration of Helsinki for Human Research. In the 2015/16 epidemic season over 8300 specimens were tested. The material consisted of nasal and throat swabs and bronchoalveolar lavage fluid (BALF). Clinical specimens were collected from week 40/2015 to week 34/2016 (October 1/2015 – August 28/2016).

2.2 Extraction of Viral RNA

Viral RNA was isolated using a Maxwell 16 Viral Total Nucleic Acid Purification Kit (Promega Corporation; Madison, WI) from 200 μL of clinical samples in phosphate-buffered saline (PBS) in accordance with the manufacturer's instructions for low elution volume cartridges. The RNA was eluted with 50 μL of RNase-free water.

2.3 Real-Time Quantitative Reverse Transcription Polymerase Chain Reaction (Q-RT-PCR)

The Q-RT-PCR was used for a molecular examination of the presence of influenza viral material. The analysis was carried out using a Light Thermocycler 2.0 System (Roche Diagnostics; Rotkreuz, Switzerland). The reaction proceeded

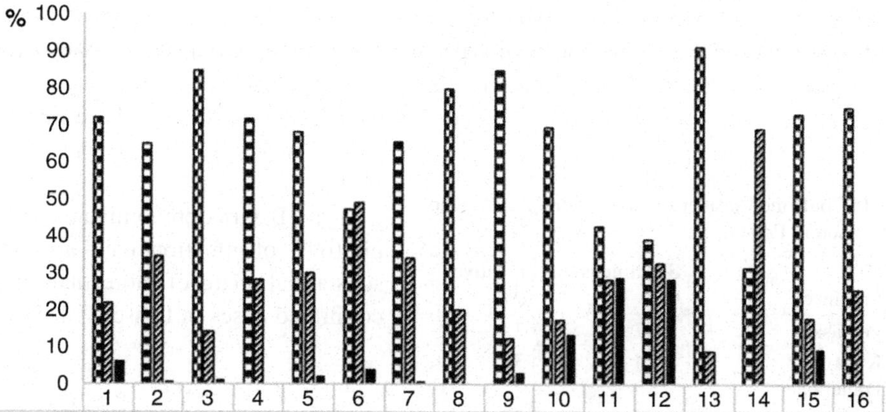

	1	2	3	4	5	6	7	8	9	10	11	12	13	14	15	16
Influenza A (%)	72	65	85	72	68	47	65	80	85	69	43	39	91	31	73	75
Influenza B (%)	22	34	14	28	30	49	34	20	12	17	28	32	9	69	18	25
ILI (%)	6	1	1	0	2	4	1	0	3	14	29	29	0	0	9	0

Influenza A (%) Influenza B (%) ILI (%)

Fig. 1 Percentage distribution of influenza and influenza-like infections in the Polish provinces in the 2015/16 epidemic season. The province number, placed under each set of bars, corresponds to the numbered names of provinces provided in the tables

in capillaries of 20 μL volume. Primers and probes necessary to carry out the reaction were obtained through the Influenza Reagent Resource (IRR) program from the US Center for Disease Control (CDC) The reaction mixture contained $MgSO_4$, reaction buffer - bovine serum albumin (BSA), RNase free water, and a SuperScript® III/platinum *Taq* mix (Invitrogen Life Technologies-Thermo Fisher Scientific; Carlsbad, CA), with the addition of 5 μL of the previously isolated RNA for each sample. The positive control constituted the RNA strains that were the following vaccine components for the 2015/16 epidemic season/2016: A/H1N1/pdm09 (A/California/7/2009), A/H3N2/Switzerland/9715293/2013, and B/Phuket/3073/2013. As for the negative, RNase-free water was used. Before the start of amplification, the RNA was rewritten to obtain cDNA, using the enzyme reverse transcriptase at 50 °C for 30 min. Then, samples were analyzed as follows: initialization at 95 °C for 2 min and 45 cycles of amplification consisting of denaturation at 95 °C for 15 s, annealing at 55 °C for 30 s, and elongation at 72 °C for 2 min.

3 Results

3.1 Influenza

In the 2015/2016 epidemic season in Poland over eight thousand five hundred specimens for influenza and influenza-like viruses were examined. The presence of influenza virus was found in over three thousand cases. The number of confirmed cases and share of various types of virus are presented in Fig. 1.

The majority of confirmations of influenza were found in the Pomeranian (72%), Masovian (62%), and Subcarpathian (55%) provinces. In other provinces, influenza was confirmed in less than half specimens investigated (Fig. 1 and Table 1).

We further found that the predominant subtype of influenza A virus in the 2015/16 season was A/H1N1/pdm09 that was detected in high percentage of influenza infection in several provinces, such as Pomerania (80%), Podlaskie (76%), Subcarpathia (74%), Lubuskie (71%), Silesia, and Warmia-Masuria (69%), with a smaller contribution from other provinces. In two provinces, Swietokrzyskie and Greater Poland, there was an equal frequency of type A and B viruses. Subtype A/H3N2/ accounted for one percent in the Kuyavian-Pomeranian and Masovian provinces. Apart from the two subtypes, there were unsubtyped cases of influenza infection (60%). These infections occurred most frequently in the province of Opole (Table 2).

Type B virus predominated in West Pomerania (69% of infections) and also accounted for a substantial share of more than one third of the confirmed cases in Lesser Poland (Table 2).

Table 1 Confirmed influenza cases in the 2015/16 epidemic season in Poland

	Province	Specimens (n)	Positive (%)
1	Podlaskie	137	21
2	Kuyavia-Pomerania	414	37
3	Pomerania	97	72
4	Lubuskie	243	32
5	Silesia	815	39
6	Swietokrzyskie	312	38
7	Lesser Poland	1502	38
8	Lublin	454	45
9	Lodz	154	41
10	Warmia–Masuria	386	25
11	Opole	61	8
12	Greater Poland	1515	20
13	Subcarpathia	571	55
14	West Pomerania	921	35
15	Masovia	681	62
16	Lower Silesia	279	40
	Total	8542	37

3.2 Influenza-Like Illness

Influenza-like infections (ILI) accounted for over 20% of all infections in the Opole and Greater Poland provinces. In other regions of the country these infections were rather sporadic (Fig. 1). The respiratory syncytial virus was the most common cause of ILI. Other viruses, such as human metapneumovirus, adenovirus, coronavirus 229E/NL63, parainfluenzavirus 1-3, coronavirus OC43, and rhinovirus A/B were identified in individual cases.

Table 2 Percentage distribution (%) of influenza virus infections in the Polish provinces in the 2015/16 epidemic season

	Province	A/H1N1/ (%)	A/H1N1/pdm09 (%)	A/H3N2/ (%)	A unsubtyped (%)	B (%)
1	Podlaskie	0	76	0	3	21
2	Kuyavia-Pomerania	0	52	1	13	34
3	Pomerania	0	80	0	6	14
4	Lubuskie	1	71	0	0	28
5	Silesia	0	69	0	0	31
6	Swietokrzyskie	2	3	0	44	51
7	Lesser Poland	0	35	0	31	34
8	Lublin	0	55	0	24	21
9	Lodz	0	62	0	25	13
10	Warmia–Masuria	1	69	0	10	20
11	Opole	0	0	0	60	40
12	Greater Poland	0	52	0	2	46
13	Subcarpathia	0	74	0	17	9
14	West Pomerania	2	21	0	8	69
15	Masovia	4	51	1	1%	25
16	Lower Silesia	0	67	0	8	25

4 Discussion

In the 2015/16 epidemic season in Poland, over three thousand respiratory virus infections were confirmed. There were 3,864,731 cases and suspected cases of influenza and influenza-like illness, 15,312 hospitalizations, and 140 deaths (NIPH-NIH 2016). Most cases and suspected cases were reported in the Pomeranian (765,296), Masovian (689,370), and Greater Poland (546,357) provinces as shown in Fig. 1. The highest incidence per 100,000 population was in Pomerania (4383.7) and Greater Poland (2086.2). The predominant subtype A/H1N1/pdm09 accounted for over 70% of confirmed cases of influenza infections. It should be noted that the WHO provides that each year from 330 million to 1.58 billion people worldwide suffer from influenza and influenza-like infections (WHO 2015).

A high percentage of infection by subtype A/H1N1/pdm09 may stem from a low level of vaccination coverage in Poland amounting to about 3.6% of the population (Brydak 2016). This subtype was recommended as a component of influenza vaccine in the 2015/16 epidemic season (WHO 2015) and it also is going to be included into the 2016/17 vaccine (WHO 2016). Influenza virus type B was the most frequent in the West Pomeranian and Swietokrzyskie provinces, accounting for more than half of the confirmed infections (Table 1). A similar number of cases of influenza B and influenza-like infections was observed in the Opole and Greater Poland provinces, amounting to about 30% each. The majority of deaths occurred in the region of Silesia (33), in which the most frequently detected virus was influenza type A. There also were 25 deaths reported in Greater Poland, where virus type A (39%), B (32%), and viruses causing influenza-like illness (29%) were at a grossly comparable level. The WHO estimates that 1 million people die each year due to influenza infection (Brydak 2012).

The present virological data indicate the predominance of infection with influenza virus type A in the eastern Polish provinces: Subcarpathian (91%), Lublin (80%), and Podlaskie (74%), lying near the border with Russia. These infection may be related to border trafficking between the two countries as type A virus was confirmed as predominant from week 3 of 2016 in Russia (Kimissarov et al. 2016). However, predominance of type A virus in 2015/16 was also confirmed in Slovakia, Ukraine, Belarus, and Lithuania (Flu News Europe 2016).

A decline in the number of confirmed cases of influenza virus subtype A/H3N2/, which accounted for only 1% in the Kuyavian-Pomerianian and Masovian provinces, was observed in 2015/16 compared with that in 2014/15 (Bednarska et al. 2015); a situation similar to that also noted in Russia where this subtype caused only sporadic infections (Kimissarov et al. 2016). In contrast, predominance of influenza virus type B was observed the West Pomerania that lies west of the provinces above mentioned near Germany where type B virus was found to predominate in the 2015/16 influenza season (Flu News Europe 2016). Therefore, it seems that the regional differences in the type of influenza virus causing infections in a given season have in all probability to do with the epidemiological situation in neighboring countries.

Acknowledgements This work was funded by NIPH-NIH thematic subject 5/EM.1. The authors would like to acknowledge physicians and employees of VSESs participating in Sentinel and non-Sentinel programs for their input into the influenza surveillance in Poland.

Conflicts of Interest The authors declare no conflicts of interests in relation to this article.

References

Bednarska K, Hallmann-Szelińska E, Kondratiuk K, Brydak LB (2015) Evaluation of the activity of influenza and influenza-like viruses in the epidemic season 2013/2014. Adv Exp Med Biol 857:1–7

Brydak LB (2012) Influenza – an age old problem. Hygeia Public Health 47(1):1–7

Brydak LB (2016) Prophylaxis of influenza in general practice. Doctor Guide 1:13–15. (Article in Polish)

Flu News Europe (2016) https://flunewseurope.org. Accessed July 2016

Hallmann-Szelińska E, Bednarska K, Korczyńska M, Paradowska-Stankiewicz I, Brydak LB (2016a) Virological characteristics of the 2014/2015 influenza season based on molecular analysis of biological material derived from I-MOVE study. Adv Exp Med Biol 921:81–85

Hallmann-Szelińska E, Bednarska K, Kondratiuk K, Rabczenko D, Brydak LB (2016b) Viral infections in children in the 2014/2015 epidemic season in Poland. Adv Exp Med Biol 912:51–56

Kimissarov A, Fadeev A, Sergeeva M, Petrov S, Sintsova K, Egorova A, Pisareva M, Buzitskaya Z, Musaeva T, Danilenko D, Konovalova N, Petrova P, Stolyarov K, Smorodintseva E, Burtseva E, Krasnoslobodtsev K, Kirillova L, Eropkin M, Sominina A, Grudinin M (2016) Rapid spread of influenza A(H1N1)pdm09 viruses with a new set of specific mutations in the international genes in the beginning of 2015/2016 epidemic season in Moscow and Saint Petersburg (Russian Federation). Influenza Other Respir Viruses 10(4):247–253

NIPH-NIH (2016) Influenza and influenza-like illness in Poland. Available from: http://wwwold.pzh.gov.pl/oldpage/epimeld/grypa/index.htm. Accessed 22 Nov 2016

Taubenberger JK, Morens DM (2008) The pathology of influenza virus infections. Annu Rev Pathol 3:499–522

WHO (2015) http://www.who.int/wer/2015/wer9011.pdf?ua=1. Accessed 20 July 2016

WHO (2016) http://www.who.int/influenza/vaccines/virus/recommendations/2016_17_north/en/. Accessed 20 July 2016

Advs Exp. Medicine, Biology - Neuroscience and Respiration (2017) 32: 7–16
DOI 10.1007/5584_2017_1

Published online: 25 February 2017

Impact of Internal and External Factors on EBC-pH and FeNO Changes in Humans Following Challenge with Ethyl Acrylate

F. Hoffmeyer, K. Sucker, H. Berresheim, C. Monsé, B. Jettkant, A. Beine, M. Raulf, J. Bünger, and T. Brüning

Abstract

Acute effects of ethyl acrylate exposure at 5 ppm for 4 h include changes of pH in exhaled breath condensate (EBC-pH) and exhaled nitric oxide (FeNO). So far, few data have been reported for atopic persons or the impact of the exposure conditions on biomarkers, e.g., constant versus variable application of irritants. Nine atopic and eighteen healthy volunteers without bronchial hyperresponsiveness were exposed for 4 h to ethyl acrylate concentrations of 0.05 ppm (sham), 5 ppm (constant concentration), and 0–10 ppm (variable, mean concentration of 5 ppm) in an exposure laboratory. A positive atopic status was defined according to specific IgE concentrations to common inhalant allergens (sx1 $\geq$ 0.35 kU/L). Biomarker levels were assessed before and after challenge and adjusted for levels after sham exposure (net response). Ethyl acrylate at constant, but not at variable concentrations induced a significant change in the net responses of EBC pH and FeNO. Concerning FeNO, this could be observed only for atopic persons. The changes of biomarker levels were related to their baseline values. Biomarker responses to challenge with ethyl acrylate may be influenced by the patterns of application as well as baseline airway inflammation and atopic status of the volunteers.

Keywords

Atopy • Ethyl acrylate • Exhaled breath condensate • Exhaled nitric oxide • Exposure profile • Irritant challenge • Susceptibility

1 Introduction

Ethyl acrylate is an ester of acrylic acid and an air pollutant in workplaces with polymer production (e.g., paper, textile, and leather industries). It has

F. Hoffmeyer (✉), K. Sucker, H. Berresheim, C. Monsé, B. Jettkant, A. Beine, M. Raulf, J. Bünger, and T. Brüning
Institute for Prevention and Occupational Medicine of the German Social Accident Insurance, Institute of the Ruhr-Universität Bochum (IPA), Bochum, Germany
e-mail: hoffmeyer@ipa-dguv.de

a characteristic unpleasant pungent odor and is irritating to mucosal membranes. Odor threshold of ethyl acrylate is typically below its threshold for irritation. These thresholds were determined at 0.0066 ppb and 4.15 ppm, respectively (van Thriel et al. 2006). Ethyl acrylate is a liquid and becomes a soluble gas at room temperature. In contact with water it forms acrylic acid and ethanol, which accounts for its irritancy to mucous membranes. The first effect induced by ethyl acrylate is irritation of eyes and upper respiratory tract (Arts et al. 2006). Inflammatory responses in the lower respiratory tract are characterized by the release of reactive oxygen species, proteolytic enzymes, and mediators. They are linked to acidification in several diseases (Kostikas et al. 2002).

Information on the composition of airway lining fluid, including pH, can be non-invasively provided by analyzing exhaled breath condensate (EBC) (Hoffmeyer et al. 2009). EBC-pH is an integrative measure of acids to bases ratio which apart from an inflammatory response can also be affected by physicochemical properties of inhaled constituents (Davis et al. 2013). Another non-invasive marker for the measurement of airway inflammation is exhaled nitric oxide (FeNO) (Kharitonov et al. 1997). FeNO is a useful inflammation marker of exposure to gases and organic solvents (Maniscalco et al. 2004). In addition to sensory irritation of eyes and nose, we observed in our previous work a net decrease of EBC-pH, but no change of FeNO, calculated on the mean response after acute challenge with ethyl acrylate at 5 ppm for 4 h (Hoffmeyer et al. 2016).

Subjects' characteristics contribute to variability of effects observed after challenge with irritants. Some of the known and postulated factors include age, gender, body size, exercise, atopy, airway responsiveness, smoking, and pre-existing respiratory disease (Utell and Frampton 2000). Also, the circumstances of exposure might influence a particular response. In this line, the actual dose at the target side, e.g., lung epithelium or the pattern of delivery influences the effects. Effective dose concepts suggest that the impairment of lung function after ozone exposure could be described as a function of ozone concentration (C), minute ventilation (V), and duration of exposure (T) (Adams et al. 1981). However, considerable variability in the magnitude of response remains unexplained when referring to a CVT model (McDonnell et al. 2010). This variability hinders the ability to predict different responses to equivalent doses of ozone inhaled in different exposure patterns. In addition, multiple effects under investigation might be unequally influenced by susceptibility. For instance, ozone-induced decrements in lung function do not correlate with measures of inflammation (Balmes et al. 1996). In stable asthmatics, an increase in markers of neutrophilic inflammation and oxidative stress can be demonstrated after ozone challenge despite missing functional impairment (Vagaggini et al. 2010).

Data on the impact of internal and external factors on EBC-pH and FeNO changes in humans following challenge with ethyl acrylate are lacking. With the above-mentioned considerations on irritants in mind, we examined responses in atopic and non-atopic subjects. In addition, factors related to the exposure condition were studied. Ethyl acrylate was delivered in a constant and variable pattern, both characterized by an equivalent mean concentration of 5 ppm.

2 Methods

The study was approved by a local Ethics Committee of the Ruhr University in Bochum. All participants gave written informed consent and received financial compensation for their participation. The protocol was created in accordance with the Declaration of Helsinki for Human Reserch.

2.1 Subjects

Twenty seven volunteers (15 female, 12 male) were recruited by advertisement and answered a health questionnaire. Before exposures, medical

Table 1 Subject characteristics

	Non-atopic	Atopic	p
Gender; F/M (n)	12/6	3/6	0.114
Age (year)	25 (23–27)	24 (23–27)	0.671
BMI (kg/m^2)	21.0 (20.0–23.3)	23.0 (21.5–24.5)	0.159
sx1 (kU/L)	0.10 (0.09–0.11)	4.31 (1.3–28.7)	<0.0001
FEV_1 (%pred$_{GLI}$)	95.5 (90.3–103.7)	97.7 (95.0–106.8)	0.287
z-score	−0.38 (−0.82; 0.31)	−0.20 (−0.43; 0.57)	0.294
FVC (%pred$_{GLI}$)	97.3 (91.3–108.0)	103.0 (97.3–110.2)	0.185
z-score	−0.23 (−0.72; 0.68)	0.23 (−0.23; 0.85)	0.178
FEV_1/FVC (%pred$_{GLI}$)	97.3 (95.5–101.8)	97.0 (94.4–99.6)	0.447
z-score	−0.42 (−0.74; 0.04)	−0.48 (−0.78; −0.04)	0.878

Continuous variables with median and inter-quartile range (IQR);. *F* female, *M* male, *BMI* body mass index, *sx1* IgE antibodies to a mixture of ubiquitous allergens (atopy screen), *FEV_1* forced expiratory volume in 1 s, *FVC* forced vital capacity

examination, analyses of blood and urine, and lung function were performed. The volunteers were healthy non-smokers without bronchial hyperresponsiveness in the methacholine challenge test and lung function values were within normal limits. Predicted values of lung function parameters and respective z-scores were derived from healthy non-smoking Caucasian subjects collected by the Global Lung Initiative (GLI) (Quanjer et al. 2012). Smoking habits were assessed by face-to-face interviews and validated by COHb (carboxyhemoglobin) and the nicotine metabolite cotinine in urine as previously described (Hoffmeyer et al. 2015a). Atopy was classified according to specific IgE concentrations to common inhalant allergens (sx1 Phadiatop; ThermoFisher Phadia AB, Uppsala, Sweden). A positive atopic status was assumed in case of sx1 $\geq$ 0.35 kU/L and 9 out of the 27 participants had serologically verified atopy. Descriptive data on the 18 non-atopic and 9 atopic participants are summarized in Table 1.

2.2 Exposure

Volunteers were in random order exposed to ethyl acrylate concentrations of 0.05 ppm (sham), 5 ppm (constant concentration), or 0–10 ppm (variable, mean concentration of 5 ppm). During the variable wave-form exposure, the concentration was increased from 0–10 ppm and then reduced back to 0 ppm each hour (Fig. 1). The exposures were performed for 4 h at rest in an exposure laboratory (ExpoLab) with up to four volunteers at a time as recently published (Hoffmeyer et al. 2016). Details of the ExpoLab have been previously reported (Monsé et al. 2012). A mass spectroscopy in chemical ionization mode (Model Airsense; MS4-Analysentechnik GmbH, Rockenberg, Germany) was used to determine ethyl acrylate concentrations every 2 s in the breathing zone.

2.3 Effect Markers

FeNO and EBC-pH were used for effect assessment as previously described (Hoffmeyer et al. 2016). Briefly, FeNO was measured following the American Thoracic Society and European Respiratory Society recommendations (ATS/ERS 2005) using a portable electrochemical analyzer (NIOX Mino; Aerocrine, Solna, Sweden). EBC was sampled following general methodological recommendations (Horváth et al. 2005) using a temperature-controlled device (Turbo DECCS; Medivac, Parma, Italy). Collection was performed during tidal breathing for 10 min at a constant device's temperature of −5 °C. As the partial pressure of CO_2 (pCO_2) is the most important confounder of pH measurement in EBC samples, we adjusted pH to a pCO_2

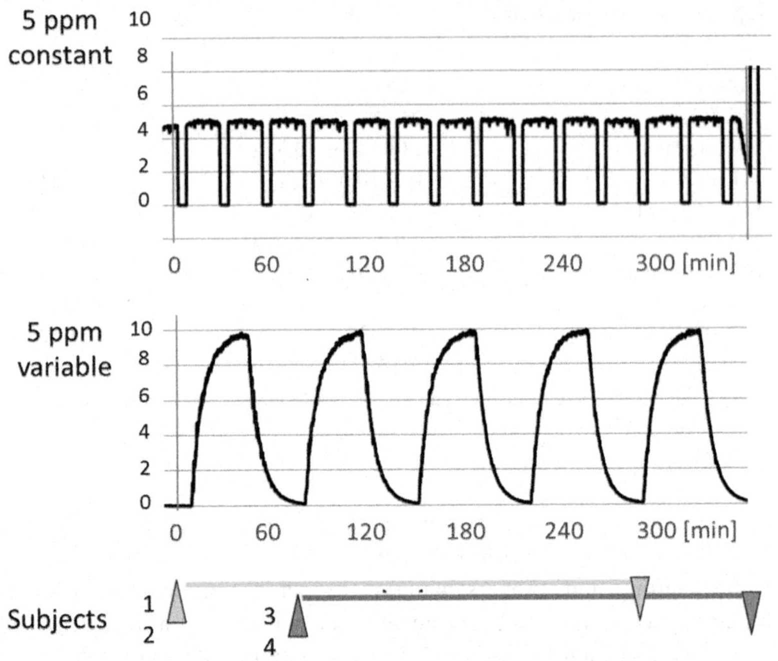

Fig. 1 Experimental design. Subjects were exposed to a sham condition (*not shown*) and ethyl acrylate in a mean concentration of 5 ppm applied either *via* constant (*top*) or variable profile (*bottom*). Drops to zero concentration (*top*) are artificial and result from switching analyses between the four breathing zones. Due to time-consuming measurements before and after exposure, subjects entered and left the ExpoLab in pairs

of 5.33 kPa as previously reported (Kullmann et al. 2007; Hoffmeyer et al. 2015b). Simultaneous measurements of pH and pCO_2 were done with a blood gas analyzer (ABL800; Radiometer GmbH, Willich, Germany). One participant was unavailable for EBC sampling after sham exposure; thus he was not included in this part of results.

2.4 Data Analysis

Changes of effect markers were normalized to the respective baseline measurements for each exposure condition ((post-pre/pre)*100, Δ%). Further, effects observed at a mean concentration of 5 ppm (either applied in a constant or variable profile) were adjusted by subtracting the results obtained after sham exposure (Δ5 ppb% -Δsham %, net-Δ%). The D'Agostino and Person omnibus normality test was used to assess the value distribution. Differences in exposure conditions were analyzed by the Wilcoxon matched-pairs signed-rank test or paired t-test as appropriate. Comparison of non-atopic and atopic volunteers was based on the Mann Whitney test or unpaired *t*-test as appropriate. The Spearman rank or Pearson correlation coefficient was used for analysis of correlation as appropriate. For all tests a significance level of <0.05 was chosen. Continuous variables are depicted with median and interquartile range (IQR; 25th, 75th percentile). Data were analyzed and visualized by using GraphPad Prism (Version 5.01, GraphPad Software, San Diego, CA).

3 Results

3.1 Effect Markers and Exposure Conditions

Two challenge conditions (constant and variable) were designed resulting in an equivalent total amount of ethyl acrylate delivered during 4 h. The effects were also assessed following an identical sham exposure. Table 2 summarizes the results for the three exposures. There were no differences in the baseline levels of either FeNO ($p = 0.685$) or pH ($p = 0.169$) among the three conditions.

Table 2 Group median effect markers after exposure (post) compared to baseline (pre) for the three challenge protocols

		Sham	Constant	Variable
FeNO (ppb)	pre	16 (11–20)	14 (12–20)	14 (11–18)
	post	13 (11–18)	12 (11–19)	14 (10–15)
	p	<0.001	0.153	<0.001
pH	pre	5.925 (5.831–5.973)	5.966 (5.874–6.032)	5.924 (5.845–5.968)
	post	6.051 (5.980–6.084)	6.014 (5.941–6.078)	6.017 (5.961–6.111)
	p	<0.001	<0.001	<0.001

Median and inter-quartile range (IQR)

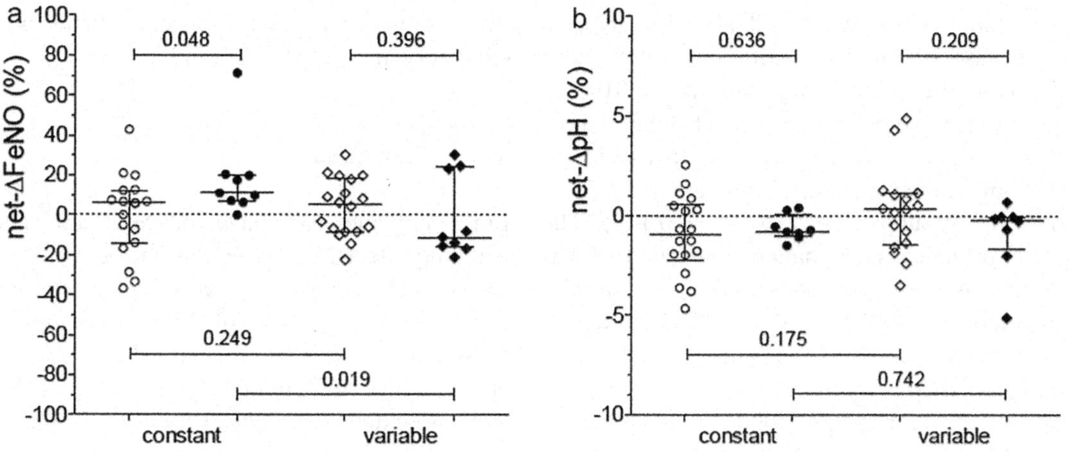

Fig. 2 Net responses of FeNO (**a**) and pH (**b**) following either constant or variable ethyl acrylate challenge. Results are adjusted for sham exposure and stratified according to non-atopics (*open symbols*) and atopics (*solid symbols*)

Compared with the baseline level, FeNO values decreased after sham condition and following the constant and variable ethyl acrylate challenges. The median FeNO changes referring to all volunteers were −11.8% (IQR −21.4, 0.0), −7.7% (IQR −18.1, 0.0), and −12.5% (IQR −18.2, −0.1), respectively. After applying the variable profile, the FeNO decline was statistically significant despite an unchanged median level. FeNO changes in response to the two ethyl acrylate challenges were not different from the sham condition (constant *vs.* sham; $p = 0.164$, variable *vs.* sham; $p = 0.475$) and demonstrated no mean difference to each other (constant *vs.* variable; $p = 0.287$).

In contrast to FeNO, pH increased after all exposure conditions. With respect to all volunteers, the changes following sham condition, constant, and variable ethyl acrylate challenge compared with the baseline levels were significant (each $p < 0.001$), demonstrating an increase of 1.83% (IQR 1.36, 2.62), 1.21% (IQR 0.44, 2.01), and 1.27% (IQR 0.48, 2.69), respectively. The increase after ethyl acrylate challenge compared with the increase after sham condition was significantly lower when applying a constant concentration (constant *vs.* sham; $p = 0.015$, variable *vs.* sham; $p = 0.554$).

3.2 Effect Markers and Atopic Background

Changes in the biomarkers following either constant or variable ethyl acrylate exposures adjusted for the sham condition (net-response) and stratified according to the atopic status are illustrated in Fig. 2. No significant net-effects on FeNO (net-ΔFeNO) of variable ethyl acrylate exposure was observed (non-atopic: 5.4%; $p = 0.269$; atopic: −11.1%; $p = 0.906$). When ethyl acrylate was administered in a constant manner, a significant increase in net-ΔFeNO was revealed for atopic but not non-atopic volunteers (non-atopic: 5.8%; $p = 0.999$, atopic: 11.1%; $p = 0.032$). None of the atopic volunteers demonstrated a net-ΔFeNO decline after constant challenge (Fig. 2a). The net-ΔFeNO increase in the atopics after constant exposure was significantly different from the effect in non-atopic volunteers ($p = 0.048$). The changes were also significantly different from those observed after variable exposure ($p = 0.019$).

In the variable ethyl acrylate challenge, there was no significant change in the sham-adjusted EBC-pH values (net-ΔpH; non-atopic: 0.30%; $p = 0.759$, atopic: −0.23%; $p = 0.178$). The effect of the constant challenge on net-ΔpH was a decrease in both non-atopics and atopics (non-atopic: −0.98%; $p = 0.052$, atopic: −0.80%; $p = 0.028$). The magnitude of decrease was similar ($p = 0.636$, Fig. 2b). No significant differences in net-ΔpH following constant or variable ethyl acrylate challenge were observed either in non-atopics (constant: −0.98% *vs.* variable: 0.30%; $p = 0.175$) or atopics (constant: −0.80% *vs.* variable: −0.23%; $p = 0.742$). No influence of the order of the three challenge conditions or of the gender on biomarker responses could be identified (data not shown).

3.3 Effect Markers and Baseline Levels

All individual measurements of FeNO before and after each of the three challenge conditions were significantly correlated with each other ($p < 0.001$). Also, pH values before and after sham and constant ethyl acrylate exposure were significantly correlated ($p < 0.001$), whereas the pH values after variable ethyl acrylate were not correlated to the baseline values ($p = 0.196$).

Correlations between the individual baseline values and the magnitude of change following the exposure protocols are shown in Table 3. Concerning FeNO, the only correlation could be revealed in case of atopics exposed to ethyl acrylate in a constant manner ($r = 0.758$; $p = 0.018$; Fig. 3a). Except for atopics exposed to ethyl acrylate in a variable manner ($p = 0.208$), correlations could be observed between changes of pH following exposures and the respective pH values at baseline (Table 3). Increases following exposures showed significant inverse correlations to the pH value before exposure; i.e., the increases were stronger for lower baseline pH values. As an example, the regression lines for the results after constant challenge with ethyl acrylate are presented in Fig. 3b.

4 Discussion

Ethyl acrylate can have odorous and also irritating effects. In addition to sensory irritation of eyes and nose, we have recently demonstrated a net decrease of EBC-pH, but no change of FeNO, in humans after acute challenge with ethyl acrylate at 5 ppm (Hoffmeyer et al. 2016). Now, we extend these observations regarding the internal and external factors that might modulate EBC-pH and FeNO following challenge with ethyl acrylate for 4 h. Concerning the evaluation of adverse health effects, susceptible groups in the population like atopic subjects have to be considered (Utell and Frampton 2000). Moreover, like in risk assessment of air pollution (Adams 2003), experimental simulations of occupational risk assessment need to employ concentration profiles that mimic those encountered during a working shift. Beside a constant background level of a substance, also variable concentrations, e.g., due to specific operating processes, have to be considered in the working area. In the present study, two types of exposure profile were applied. A usual square-wave profile with ethyl acrylate concentration maintained constant at 5 ppm, and a wave-form in which the concentration was increased from zero to 10 ppm and then back to zero level during one hour resulting in four exposure cycles with a mean concentration of 5 ppm for each participant. The maximum concentration of 10 ppm ethyl acrylate was chosen in accordance with the

Table 3 Relationship between changes of effect marker and baseline values for each challenge protocol (Δ) and adjusted for sham results (net−Δ)

			Δ			net−Δ	
			Sham	Constant	Variable	Constant	Variable
FeNO (ppb)	All	r	−0.278	−0.051	−0.013	−0.097	−0.015
		p	0.160	0.801	0.950	0.629	0.941
	Non-atopic	r	−0.272	−0.212	−0.047	−0.088	−0.007
		p	0.273	0.399	0.852	0.728	0.977
	Atopic	r	−0.524	0.758	0.056	0.682	0.074
		p	0.148	0.018	0.885	0.043	0.839
pH	All	r	−0.692	−0.565	−0.534	−0.357	−0.666
		p	<0.001	0.002	0.006	0.087	<0.001
	Non-atopic	r	−0.754	−0.621	−0.635	−0.459	−0.806
		p	<0.001	0.006	0.008	0.074	<0.001
	Atopic	r	−0.821	−0.906	−0.464	−0.609	−0.310
		p	0.013	<0.001	0.208	0.109	0.462

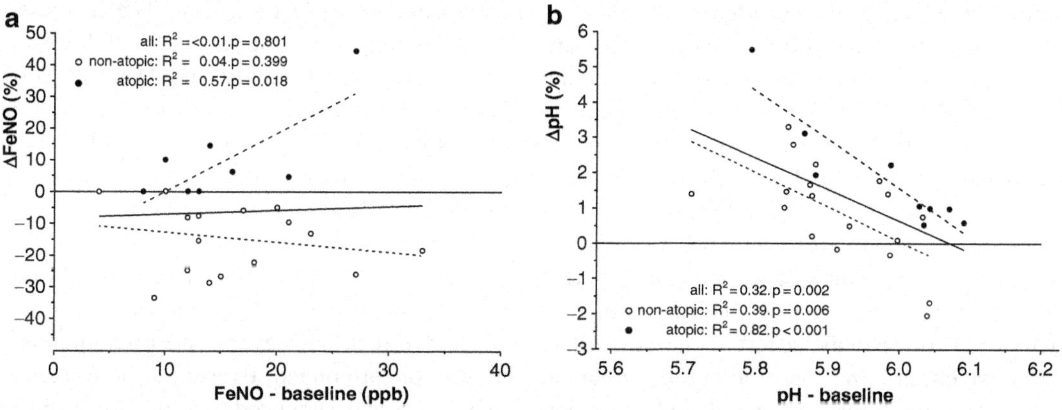

Fig. 3 Relationship between FeNO (**a**) and pH (**b**) changes following constant ethyl acrylate challenge and their respective baseline values. Results are stratified according to non-atopics (*open circles and dotted line*) and atopics (*solid circles and hashed line*). Regression analyses are also shown for the total group (*line*)

currently adjusted short-term doubling threshold limit value at the time of the study (MAK 2012). Despite the fact that equivalent total inhaled doses in both exposure profiles were delivered, we observed differing responses. The effects were assessed by adjusting responses observed after a particular concentration to those observed in an identical sham exposure as recommended (Bartoli et al. 2013; Horstman et al. 1995). First of all, we confirm our previous results on EBC-pH and FeNO changes after sham exposure regarding the healthy non-atopic subjects following a challenge with ethyl acrylate at a constant concentration of 5 ppm (Hoffmeyer et al. 2016). As in other studies of the kind we focused on younger, non-smoking participants to avoid potential problems and confounding biomarker responses associated with age-related morbidity and use of medication (Koczulla et al. 2010; Riediker and Danuser 2007; Kharitonov et al. 1995). Our investigation took place out of pollen season and all atopic participants did not show any symptoms, lung function impairment, or airway hyperresponsiveness. We did not observe increased baseline FeNO in the atopic subjects, which is being present in subjects with clinically apparent atopy (Tossa et al. 2010; De Zotti and Bovenzi 2000).

Data from the literature suggest an exposure response model being a function of irritant concentration (C), minute ventilation (V), and exposure time (T) (McDonnell et al. 2010). After adjustment for the sham condition, we observed significant effects only after exposure to a constant ethyl acrylate concentration and in case of FeNO only in atopic participants. In our experimental setting, ethyl acrylate concentration was handled in a way to deliver the same overall dose during the challenge duration of 4 h. However, the profile with changing concentration included periods of lower and down to zero decreasing content, which may have reversed the effect induced by prior exposure.

Gases, due to their physicochemical properties, promote biological and neural reflex responses which could be interlinked (Brüning et al. 2014). Individuals often report ocular sensation disorders in addition to sensations coming from the upper and lower airways on exposures to volatile irritants. Eye irritation can be estimated by the eye blink rate. During a variable profile of exposure to an equivalent mean concentration of 5 ppm ethyl acrylate, a significant increase in eye blinking has been reported (Blaszkewicz et al. 2010). Thus, continuous activation of airway afferent nerves alters responsiveness in an adaptive manner. Concerning the airways, our previous study has indicated that tissue irritation or inflammation is triggered in response to ethyl acrylate-induced changes in FeNO and pH (Hoffmeyer et al. 2016). In this respect, it has been shown that increasing pH decreases FeNO by buffering airway acid (Gaston et al. 2006). In the present study, declining FeNO values after constant ethyl acrylate exposure were associated with increasing pH in non-atopics. The atopic volunteers did not demonstrate this inverse association. In fact, we could demonstrate a significant increase in FeNO relating to the baseline level, in this subgroup. FeNO is foremost considered as a measure of eosinophilic inflammation in asthma (Ricciardolo et al. 2015). However, there are also reports indicating that FeNO might be a useful marker of exposure to dust and gases (Hoffmeyer et al. 2015a; Maniscalco et al. 2004; Ulvestad et al. 2001). The present finding of a predominant increase in FeNO in atopic subjects exposed to ethyl acrylate suggests that atopy is associated with upregulation of inflammatory mechanisms and a greater response to ethyl acrylate. Concerning sulfur dioxide, Nowak et al. (1997) have reported that atopy, based on skin-prick results to common allergens, is a predictor of positive lung-function response. Nasal reactivity to irritant provocation is also influenced by allergy status (Shusterman et al. 2003).

Finally, our findings are in qualitative agreement with experiments supporting the concept that the baseline level of airway inflammation indicators may be a determinant of responses to stressors. In this respect, increased EBC-pH could be observed after moderate exercise corresponding to fast walking. The increase in pH was strongest for the subjects with lowest pH before exercise (Riediker and Danuser 2007). Baseline FEV_1 and FEV_1/FVC (%predicted) are indicators of the severity of asthma and are lower in subjects demonstrating ozone-induced lung impairment (Horstman et al. 1995).

In conclusion, we assume that rather small changes in pH and FeNO in response to ethyl acrylate exposure do not represent an acute adverse effect. However, under conditions of higher or prolonged exposure these changes may become pronounced and indicate a sensory or tissue irritation. Ethyl acrylate at constant, but not at variable concentration induces significant changes in the net response of EBC-pH and FeNO. Concerning FeNO, this could be observed only for atopic persons. Therefore, subjects of clinical studies should be carefully characterized to identify potential susceptible subgroups and to avoid misinterpretation of responses.

Acknowledgements We gratefully acknowledge the technicians of IPA Jennifer Gili, Ursula Meurer, Anja Molkenthin, Melanie Ulbrich, and Susann Widmer. The statements and conclusions in this article are those of the authors and not necessarily those of the German Social Accident Insurance.

Competing Interests The authors declare that they have no competing interests that might influence the results of this report.

References

Adams WC (2003) Comparison of chamber and face mask 6.6-hour exposure to 0.08 ppm ozone *via* square-wave and triangular profiles on pulmonary responses. Inhal Toxicol 15:265–281

Adams WC, Savin WM, Christo AE (1981) Detection of ozone toxicity during continuous exercise via the effective dose concept. J Appl Physiol Respir Environ Exerc Physiol 51:415–422

Arts JH, de Heer C, Woutersen RA (2006) Local effects in the respiratory tract: relevance of subjectively measured irritation for setting occupational exposure limits. Int Arch Occup Environ Health 79:283–298

ATS/ERS (2005) Recommendations for standardized procedures for the online and offline measurement of exhaled lower respiratory nitric oxide and nasal nitric oxide. American Thoracic Society European Respiratory Society. Am J Respir Crit Care Med 171 (8):912–930

Balmes JR, Chen LL, Scannell C, Tager I, Christian D, Hearne PQ, Kelly T, Aris RM (1996) Ozone-induced decrements in FEV1 and FVC do not correlate with measures of inflammation. Am J Respir Crit Care Med 153:904–909

Bartoli ML, Vagaggini B, Malagrinò L, Bacci E, Cianchetti S, Dente FL, Novelli F, Costa F, Paggiaro P (2013) Baseline airway inflammation may be a determinant of the response to ozone exposure in asthmatic patients. Inhal Toxicol 25:127–133

Blaszkewicz M, Hey K, Kiesswetter E, Kleinbeck S, Schäper M, van Thriel C (2010) Abgrenzung und Differenzierung "irritativer" und "belästigender" Effekte von Gefahrstoffen. Abschlussbericht. DGUV (Deutsche Gesetzliche Unfallversicherung). http://www.dguv.de/medien/ifa/de/pro/pro1/ff-fp0267/ff_fp0267_abschlussbericht.pdf. Accessed 10 Dec 2016

Brüning T, Bartsch R, Bolt HM, Desel H, Drexler H, Gundert-Remy U, Hartwig A, Jäckh R, Leibold E, Pallapies D, Rettenmeier AW, Schlüter G, Stropp G, Sucker K, Triebig G, Westphal G, van Thriel C (2014) Sensory irritation as a basis for setting occupational exposure limits. Arch Toxicol 88:1855–1879

Davis MD, Walsh BK, Dwyer ST, Combs C, Vehse N, Paget-Brown A, Pajewski T, Hunt JF (2013) Safety of an alkalinizing buffer designed for inhaled medications in humans. Respir Care 58:1226–1232

De Zotti R, Bovenzi M (2000) Prospective study of work related respiratory symptoms in trainee bakers. Occup Environ Med 57:58–61

Gaston B, Kelly R, Urban P, Liu L, Henderson EM, Doctor A, Teague WG, Fitzpatrick A, Erzurum S, Hunt JF (2006) Buffering airway acid decreases exhaled nitric oxide in asthma. J Allergy Clin Immunol 118:817–822

Hoffmeyer F, Raulf-Heimsoth M, Brüning T (2009) Exhaled breath condensate and airway inflammation. Curr Opin Allergy Clin Immunol 9:16–22

Hoffmeyer F, Sucker K, Monsé C, Berresheim H, Jettkant B, Rosenkranz N, Brüning T, Bünger J (2015a) Different patterns in changes of exhaled breath condensate pH and exhaled nitric oxide after ozone exposure. Adv Exp Med Biol 834:39–47

Hoffmeyer F, Berresheim H, Beine A, Sucker K, Brüning T, Bünger J (2015b) Methodological implications in pH standardization of exhaled breath condensate. J Breath Res 9(3):036003

Hoffmeyer F, Bünger J, Monsé C, Berresheim H, Jettkant B, Beine A, Brüning T, Sucker K (2016) Clinical effects, exhaled breath condensate ph and exhaled nitric oxide in humans after ethyl acrylate exposure. Adv Exp Med Biol 921:11–20

Horstman DH, Ball BA, Brown J, Gerrity T, Folinsbee LJ (1995) Comparison of pulmonary responses of asthmatic and nonasthmatic subjects performing light exercise while exposed to a low level of ozone. Toxicol Ind Health 11:369–385

Horváth I, Hunt J, Barnes PJ, Alving K, Antczak A, Baraldi E, Becher G, van Beurden WJ, Corradi M, Dekhuijzen R, Dweik RA, Dwyer T, Effros R, Erzurum S, Gaston B, Gessner C, Greening A, Ho LP, Hohlfeld J, Jöbsis Q, Laskowski D, Loukides S, Marlin D, Montuschi P, Olin AC, Redington AE, Reinhold P, van Rensen EL, Rubinstein I, Silkoff P, Toren K, Vass G, Vogelberg C, Wirtz H, ATS/ERS Task Force on Exhaled Breath Condensate (2005) Exhaled breath condensate: methodological recommendations and unresolved questions. Eur Respir J 26(3):523–548

Kharitonov SA, Robbins RA, Yates D, Keatings V, Barnes PJ (1995) Acute and chronic effects of cigarette smoking on exhaled nitric oxide. Am J Respir Crit Care Med 152:609–612

Kharitonov S, Alving K, Barnes PJ (1997) Exhaled and nasal nitric oxide measurements: recommendations. The European Respiratory Society Task Force. Eur Respir J 10:1683–1693

Koczulla AR, Noeske S, Herr C, Jörres RA, Römmelt H, Vogelmeier C, Bals R (2010) Acute and chronic effects of smoking on inflammation markers in exhaled breath condensate in current smokers. Respiration 79:61–67

Kostikas K, Papatheodorou G, Ganas K, Psathakis K, Panagou P, Loukides S (2002) pH in expired breath condensate of patients with inflammatory airway diseases. Am J Respir Crit Care Med 165:1364–1370

Kullmann T, Barta I, Lázár Z, Szili B, Barát E, Valyon M, Kollai M, Horváth I (2007) Exhaled breath condensate pH standardised for CO_2 partial pressure. Eur Respir J 29:496–501

MAK (2012) The MAK collection for occupational health and safety, Wiley Online Library, Online ISBN: 9783527600410. doi: 10.1002/3527600418

Maniscalco M, Grieco L, Galdi A, Lundberg JO, Sofia M (2004) Increase in exhaled nitric oxide in shoe and leather workers at the end of the work-shift. Occup Med (Lond) 54:404–407

McDonnell WF, Stewart PW, Smith MV (2010) Prediction of ozone-induced lung function responses in humans. Inhal Toxicol 22:160–168

Monsé C, Sucker K, van Thriel C, Broding HC, Jettkant B, Berresheim H, Wiethege T, Käfferlein H, Merget R, Bünger J, Brüning T (2012) Considerations for the design and technical setup of a human whole-body exposure chamber. Inhal Toxicol 24:99–108

Nowak D, Jörres R, Berger J, Claussen M, Magnussen H (1997) Airway responsiveness to sulfur dioxide in an adult population sample. Am J Respir Crit Care Med 156:1151–1156

Quanjer PH, Stanojevic S, Cole TJ, Baur X, Hall GL, Culver BH, Enright PL, Hankinson JL, Ip MS, Zheng J, Stocks J, ERS Global Lung Function Initiative (2012) Multi-ethnic reference values for spirometry for the 3-95-yr age range: the global lung function 2012 equations. Eur Respir J 40(6):1324–1343

Ricciardolo FL, Sorbello V, Ciprandi G (2015) FeNO as biomarker for asthma phenotyping and management. Allergy Asthma Proc 36:e1–e8

Riediker M, Danuser B (2007) Exhaled breath condensate pH is increased after moderate exercise. J Aerosol Med 20:13–18

Shusterman D, Murphy MA, Balmes J (2003) Influence of age, gender, and allergy status on nasal reactivity to inhaled chlorine. Inhal Toxicol 15:1179–1189

Tossa P, Paris C, Zmirou-Navier D, Demange V, Acouetey DS, Michaely JP, Bohadana A (2010) Increase in exhaled nitric oxide is associated with bronchial hyperresponsiveness among apprentices. Am J Respir Crit Care Med 182:738–744

Ulvestad B, Lund MB, Bakke B, Djupesland PG, Kongerud J, Boe J (2001) Gas and dust exposure in underground construction is associated with signs of airway inflammation. Eur Respir J 17:416–421

Utell MJ, Frampton MW (2000) Toxicologic methods: controlled human exposures. Environ Health Perspect 108(Suppl 4):605–613

Vagaggini B, Bartoli ML, Cianchetti S, Costa F, Bacci E, Dente FL, Di Franco A, Malagrinò L, Paggiaro P (2010) Increase in markers of airway inflammation after ozone exposure can be observed also in stable treated asthmatics with minimal functional response to ozone. Respir Res 11:5

van Thriel C, Schäper M, Kiesswetter E, Kleinbeck S, Juran S, Blaszkewicz M, Fricke HH, Altmann L, Berresheim H, Brüning T (2006) From chemosensory thresholds to whole body exposures-experimental approaches evaluating chemosensory effects of chemicals. Int Arch Occup Environ Health 79:308–321

Advs Exp. Medicine, Biology - Neuroscience and Respiration (2017) 32: 17–24
DOI 10.1007/5584_2017_14

Published online: 25 February 2017

Influence of Socioeconomic Factors on Self-Reported Prevalence of Allergic Diseases Among Female University Students

K. Kliś, M. Żurawiecka, A. Suder, I. Teul,
B. Borowska-Strugińska, E. Suliga, and I. Wronka

Abstract

Until recently, most studies report an increasing prevalence of allergy and asthma. The research suggests that the increase may have to do with changes in lifestyle and living conditions. This study seeks to determine the prevalence and changes in allergic diseases in relation to socioeconomic status (SES) 6 years apart. The research material consisted of data collected in two cross-sectional surveys conducted among university female students in 2009 and 2015 (respectively, 702 and 1305 subjects). The surveys evaluated the incidence of allergic conditions and socio-economic status. The occurrence of allergy was determined on the basis of answers to the questions whether the allergy and specific allergens were defined on the basis of medical work-up. The prevalence of allergic diseases increased from 14.0% to 22.3% over a 6-year period. In both cohorts, allergic diseases were more prevalent among females with high SES than with low SES. In 2009, significant differences were noted in relation to urbanization of the place of living and the number of siblings. In 2015, all socioeconomics factors significantly bore on the prevalence of allergy.

K. Kliś, M. Żurawiecka, and I. Wronka (✉)
Department of Anthropology, Institute of Zoology, Jagiellonian University, 9 Gronostajowa Street, 30–387 Krakow, Poland
e-mail: iwona.wronka@uj.edu.pl

A. Suder
Department of Anatomy, University of Physical Education, Krakow, Poland

I. Teul
Department of Anatomy, Pomeranian Medical University in Szczecin, Szczecin, Poland

B. Borowska-Strugińska
Faculty of Biology and Environmental Protection, Department of Anthropology, University of Lodz, Lodz, Poland

E. Suliga
Faculty of Medicine and Health Sciences, Department of the Prevention of Alimentary Tract Diseases, Institute of Nursing and Obstetrics, Jan Kochanowski University, Kielce, Poland

Keywords
Allergy • Education • Life style • Place of living • Siblings • Socioeconomic status • Survey • Urbanization

1 Introduction

A rise in allergic morbidity has been observed over the last few decades. A key role in the process is attributed to environmental factors. The prevalence of allergy is also influenced by genetic factors. However, a change of the gene pool is not likely to take place in such a short period of time. It is also known that etiology of allergic diseases is underlain by multiple gene functions; therefore allergy development depends on interactions between the genotype and environment (Holloway et al. 2010).

The hygiene hypothesis is one of the most popular theories explaining the widespread of allergies, mainly in highly developed countries (Okada et al. 2010; Strachan 1989). It assumes that changes in lifestyle and improvements in hygiene have caused a decrease in the prevalence of infections, and subsequently an increase in allergic morbidity. The mechanisms of this interrelation are complex. They are connected with a decreased consumption of homeostatic factors and immunoregulation, involving stimulation of various regulatory T-cell subsets and Toll-like receptors. Changes in microbiota caused by changes in lifestyle, notably articulated, for instance, in inflammatory bowel diseases, could influence the processes outlined above (Okada et al. 2010).

It is commonly known that living conditions and lifestyle are determined by people's socioeconomic status (SES). These conditions are usually assessed based the level of urbanization of the living place, parents' education, number of siblings, earnings, including money spent on living comfort, food, medical, and the knowledge on disease prevention and healthy lifestyle. Consequences of differences in SES are visible in the ontogenetic rate in children and adolescent, body build, and biological condition in children, adolescents and adults. Generally, people with high socio-economic status reach various child development stages at younger ages and are characterized by better nutrition and health. The general morbidity is usually lower in groups with high SES socio-economic status. However, a reverse tendency is observed in case of allergic diseases (Jackson et al. 2004; Cavelaars et al. 1998; Marmot et al. 1991). According to the hygiene hypothesis, factors more common in groups of low SES, such as staying in large groups of children or exposure to infectious diseases in childhood, prevent the occurrence of atopic diseases.

The literature demonstrates that a high SES increases the risk of food allergies, allergic rhinitis, and eczema in both children and parents (Jansson et al. 2014). We confirmed that in a previous study conducted among university students (Pawlińska-Chmara and Wronka 2009). However, there also are studies demonstrating that both asthma and allergic rhinitis are more often found in people of low than high status (Uphoff et al. 2015). Due to the existing controversies on the link between allergic diseases and SES, compounded by the ever changing socio-economic and cultural changes in society, in the present study we seek to determine differences in allergy prevalence with regard to the socioeconomic status in a cross sectional analysis of 2009 and 2015 periods.

2 Methods

The study was approved by a local Ethics Committee of the Jagiellonian University, in Cracow, Poland and was conducted in accord with the principles of the Declaration of Helsinki for Human Research. The material consisted of data collected in two cross-sectional surveys conducted among female university students in 2009 and 2015. In total 2007 subjects were surveyed; 702 in 2008 and 1305 in 2015. The mean age in the first cohort was 20.4 ± 1.6 years (women without allergy – 20.4 ± 1.7 years and with allergy – 20.4 ± 1.4 years) and in the second cohort it was 20.4 ± 1.7 years (women without allergy – 20.4 ± 1.6 years and with allergy – 20.4 ± 1.3 years). There were no appreciable age-difference between allergic and non-allergic subjects in either cohort. Each student responded to a questionnaire. The questions concerned the incidence of allergic conditions, SERS, age at onset of adolescence, and the menstrual cycle pattern.

The incidence of atopy was determined using the response to the question 'Have you been diagnosed with allergy on the basis of medical tests; if yes, which allergens you are allergic to?'. The SES was established by means of standard variables used in auxological studies such as: place of residence before the university or college period, parents' level of education, and the number of siblings. The place of residence was considered in three categories: village, town (up to 100,000 inhabitants), and city (more than 100,000 inhabitants). The parents' education was considered as vocational, secondary, and higher. The number of siblings was top coded at four.

On the basis of the data above outlined, a complex SES indicator was calculated. Participants were stratified into three groups of low, average, and high status. This stratification was introduced on the basis of a value of the first component in the principal component analysis that converts a set of observations of possibly correlated variables into a set of values of linearly uncorrelated components. A Chi-squared test was used to compare categorical variables. Logistic regression was used to generate odds ratios (OR) and 95% confidence intervals (CI) to assess the overall trends set by SES. A p-value of <0.05 was considered a cut-off level for statistical significance. The analyses were performed using a commercial Statistica ver. 10 packet (StatSoft Inc.; Tulsa, OK).

3 Results

The major finding was that the prevalence of allergies significantly increased with the passing years. The percentage of female students with allergies was 15.0% in 2009 and 22.3% in 2015. In both cohorts, respiratory allergies were predominant. The socio-economic factors were connected with allergy prevalence. In 2009, significant differences were observed regarding the following variables: place of living before study, number of siblings, and a socio-economic status. No significant differences were found in allergy prevalence in relation to mother and father's education, although allergy tended to be more frequent in students whose one or both parents had university education. In 2015, significant associations were found between allergy prevalence and each of the socio-economic factors analyzed (Table 1). In both years, the percentage of students suffering from allergies was twice higher in urban than rural areas. Moreover, the higher level of parents' education, the higher allergy prevalence was noted.

In the cohort examined in 2015, prevalence of allergies was higher by 7.3% than in 2009. An increase in the prevalence of allergic diseases between 2009 and 2015 was observed across all categories of SES. A minimum gain was 2.2%. The highest gains were found in the only children (14.3%) and in students whose mothers or fathers had higher education (10.5% and 14.8%, respectively). The difference in allergy prevalence between 2019 and 2015 in students from urban and the rural areas was 9.3% and 4.3%, respectively. The increases in the prevalence of allergic diseases went hand in hand with a greater level of SES (Table 2). Logistic regression analysis confirmed that risk of allergic diseases rose along with SES in both cohorts (Table 3).

4 Discussion

Allergic diseases, mainly asthma, rhinitis, and eczema, are common in both adults and children. These diseases are not a cause of high mortality,

Table 1 Prevalence of self-reported physician-confirmed allergic diseases in 2009 in relation to socio-economic factors

	Non allergic subjects; n (%)	Allergic subjects; n (%)	p	Non allergic subjects; n (%)	Allergic subjects; n (%)	p
	2009			**2015**		
Dwelling place						
Rural	276 (90.8)	28 (9.2)	**0.003**	474 (86.5)	74 (13.5)	<**0.001**
Urban	321 (80.7)	77 (19.4)		540 (71.3)	217 (28.7)	
Mother's education						
Primary	128 (89.5)	15 (10.5)	0.072	311 (87.4)	45 (12.6)	<**0.001**
Secondary	310 (86.6)	48 (13.4)		383 (78.5)	105 (21.5)	
University	159 (79.1)	42 (20.9)		320 (69.4)	141 (30.6)	
Father's education						
Primary	159 (86.0)	26 (14.1)	0.603	425 (83.7)	83 (16.3)	<**0.001**
Secondary	257 (86.0)	42 (14.1)		325 (79.1)	86 (20.9)	
University	181 (83.0)	37 (17.0)		230 (68.3)	107 (31.8)	
Number of siblings						
0	155 (79.5)	40 (20.5)	**0.033**	148 (65.2)	79 (34.8)	<**0.001**
1	237 (85.9)	39 (14.1)		487 (77.8)	139 (22.2)	
2	140 (87.0)	21 (13.0)		232 (80.6)	56 (19.4)	
3 and more	65 (92.9)	5 (7.1)		144 (89.4)	17 (10.6)	
Socio economic status						
Low	208 (92.4)	17 (7.6)	**0.001**	376 (88.3)	50 (11.7)	<**0.001**
Medium	220 (88.4)	29 (11.7)		360 (80.0)	90 (20.0)	
High	169 (74.1)	59 (25.9)		278 (64.8)	151 (35.2)	

Bold type, statistically significant difference; p-value based on Chi^2 test

although they considerably affect the quality of life. The literature demonstrates that prevalence of allergic diseases varies depending on the region and ranges from 1 to 40%. The lowest percentage is observed in Tibet and the highest is in highly developed countries (Asher et al. 2006). In Poland, the level of allergic diseases is similar to that in highly developed countries (Samolinski et al. 2009). It should be emphasized that a significant increase in the number of allergic persons has taken place in the last decades (Duggan et al. 2012; Asher et al. 2006). Analogous results were obtained in the present work.

In the International Asthma and Allergies in Childhood (ISAAC) studies that were conducted among 6–7 and 13–14-year-old children in more than 106 centers in 56 countries, the results of phase I were compared with those of phase III performed 5 years later. It has been demonstrated that there was an increase in symptom frequency of at least one of the allergic diseases in the majority of centers. The findings of the present study, in general, confirmed those above outlined, pointing to an increasing tendency in the occurrence of allergy in Poland. Some studies indicate that asthma prevalence has changed least, the prevalence of allergic diseases actually may not change anymore, although it remains at a high level of 20–30%, and there are substantial variations inter- and intra-population (Eriksson et al. 2012).

Several works have focused on a search for the cause of an increasing allergy trend. The environmental risk factors of allergies include

Table 2 Difference (Δ%) in prevalence of self-reported physician-confirmed allergic diseases between 2015–2009 in relation to socio-economics factors

	Δ%	p
Dwelling place		
Rural	4.29	0.013
Urban	9.32	<0.001
Mother's education		
Primary	2.15	0.034
Secondary	8.11	0.004
University	10.50	0.001
Father's education		
Primary	2.29	0.047
Secondary	6.87	0.008
University	14.78	<0.001
Number of siblings		
0	14.29	<0.001
1	8.07	<0.001
2	6.40	0.010
3 and more	3.42	0.029
Socio-economic status		
Low	4.18	0.016
Medium	8.35	0.002
High	9.32	0.001

p-values based on Chi^2 test

Table 3 Risk factors for self-reported physician-confirmed allergic disease in 2009 and 2015 analyzed by multiple logistic regression

Risk factor	OR (95% CI) **2009**	OR (95% CI) **2015**
Dwelling place		
Rural	1 – Ref	1 – Ref
Urban	2.23 (1.12–4.36)	2.36 (1.49–3.79)
Mother's education		
Primary	1 – Ref	1 – Ref
Secondary	1.10 (0.87–1.93)	1.61 (0.88–2.92)
University	1.40 (0.91–1.98)	1.97 (1.08–3.39)
Father's education		
Primary	1 – Ref	1 – Ref
Secondary	1.25 (0.72–2.16)	1.47 (1.10–1.96)
University	1.44 (0.68–2.01)	1.84 (0.89–3.11)
Number of siblings		
3 and more	1 – Ref	1 – Ref
2	1.03 (0.55–1.81)	0.99 (0.59–1.69)
1	1.46 (0.84–2.01)	1.61 (1.03–2.50)
0	1.97 (0.65–2.81)	2.18 (1.36–3.48)
Socio-economic status		
Low	1 – Refl	1 – Refl
Medium	1.36 (1.05–1.77)	1.92 (1.30–2.82)
High	2.03 (1.02–4.04)	2.61 (1.39–4.87)

air pollution, exposure to endotoxins and allergens at home and work, limited exposure to microbes in the early childhood, overuse of antibiotics, low fiber diet, sedentary lifestyle, and stress (Hsu and Campbell 2016; Samolinski et al. 2012; Liu 2002). All these factors are linked to the civilization advance and socio-economic status, and consequently a more frequent prevalence of allergies (Mercer et al. 2004; Chen et al. 2002; Lewis and Britton 1998; Goh et al. 1996; Williams et al. 1994) and atopies as determined by skin tests (Schäfer et al. 2001; Forastiere et al. 1997). However, some reports do not point to the presence of such a relation (Nathan et al. 1997). Discrepant data mainly refer to asthma. Mielck et al. (1996), in a review article, have argued that the test results indicating a positive relation between asthma and SES are as frequent as the ones indicating a reverse dependence, and in some studies no relation has been found. In Poland, significant socio-economic gradients in allergic disease prevalence have been previously observed (Majkowska-Wojciechowska et al. 2007) and the present study is in line with those data. A difference between the groups stratified by the SES level was greater in 2015 than in 2009. Differences due to the parents' education level were not significant in 2009 either.

According to the 'hypothesis of hygiene', an increase in allergy results from the elimination of microorganisms stimulating development of human's immune system. Children from wealthier families of higher social status tend to stay in 'sterile development conditions'. However, etiology of different allergic diseases can vary. Additionally, there exist several factors correlating with the prevalence of a particular allergic disease. The ISAAC studies have shown wide differences between medical centers, being not due to a random sampling variation, and hardly explicable by the action of only one environmental factor. For example, in a few Latin America centers and in highly developed countries a comparable high level of prevalence of allergy has been found despite significant differences in SES, living conditions, and lifestyle (Asher et al. 2006).

The environment can influence the development of allergic diseases at both individual and population levels. However, the influence seems different in developed and developing countries and concerns not only the prevalence but also symptoms of a disease. Symptoms and disease course are less severe in persons of high than low SES, which can be explained by a higher knowledge on pro-health activities and the ability to finance preventive activities or disease recurrence (Pawlińska-Chmara et al. 2013; Gehring et al. 2006; Lewis et al. 2001).

The socio-economic factors connected with risk of allergic disease include urbanization of the place of living. Urban air pollution caused by industry and high road traffic facilitates allergy, whereas high endotoxin and other microbial exposure in the early childhood resulting from living on a farm and having contact with animals is protective against allergy (Wennergren et al. 2010; Liu 2004). Likewise, lower risk is observed in people who had animals at home in childhood (Schmitz et al. 2012).

Variability of living conditions can be a problem while examining relations between SES and allergy prevalence. The conditions can significantly vary from childhood to adolescence and adulthood. It is not fully explained whether an inverse relation between the incidence of infectious diseases and allergies refers only to early childhood or to later periods of life as well (Liu 2002, 2004). That is exemplified by a relation between exposure to endotoxins and allergy occurrence. Such exposure prevents atopy in early childhood, but is the cause of on-the-job asthma while in work environment (Liu 2002; Reed and Milton 2001).

An increased number of people with allergies has been observed in many countries in Europe, America and Asia in recent years (Kim et al. 2016; Wang et al. 2016; Duggan et al. 2012; Anthracopoulos et al. 2011; Hansen et al. 2000). Allergy is often linked to westernization of lifestyle typical for highly developed countries. The latest studies show that a high and increasing level of allergic diseases is also observed in Asian and African countries with increasing SES (Lee et al. 2016). Apparently, a high call social level and improved economic status are conducive for allergic diseases (Mercer et al. 2004). It is believed that factors affecting the development of asthma and other allergies can act differently in the developed and developing countries (D'Amato et al. 2016). The key role is played by the interaction of allergy risk factors with socio-economic status (Asher et al. 2006). Studies show that allergy prevalence increases with air pollution caused mainly by road traffic (Duggan et al. 2012), overattentiveness to hygiene (Baurecht et al. 2007; Cork et al. 2006; Sherriff et al. 2002), and with a decrease in infectious illnesses. Some allergy risk factors are also connected with cultural background. The number of families with only child is growing, children stay home with mothers or nannies in early years, and start attending kindergartens at a later age with a lower number of children in groups. Children also spend less time outdoor and tend to have a limited contact with peers than they used to do in former times. Consequently, they are less exposed to factors protecting against allergies, i.e., endotoxins, microbes, and exposure to the sun that ensures a proper vitamin D synthesis (Hwang et al. 2016). Some studies underscore the possible links between the increased prevalence of allergy and climate changes, causing changes in pollen and spores content and dispersion.

The present work underlines a growing intensity of allergy in Poland. More people with allergies were observed in all different socio-economic groups over 5 years. In a group of highest SES, the difference between 2009 and 2015 was highest. It can be assumed that the prevalence of allergic diseases in Poland has not yet reached its plateau. A further increase is expected to occur mainly in people of a low SES whose living conditions and lifestyle will likely be rising. Continuous monitoring of factors influencing allergy occurrence would help keep this sort of disease in check through preventive programs and appropriate healthcare.

Conflicts of Interest The authors declare no conflicts of interest in relation to this article.

References

Anthracopoulos MB, Fouzas S, Pandiora A, Panagiotopoulou E, Liolios E, Priftis KN (2011) Prevalence trends of rhinoconjunctivitis, eczema, and atopic asthma in Greek school children: four surveys during 1991–2008. Allergy Asthma Proc 32:56–62

Asher MI, Montefort S, Björkstén B, CKW L, Strachan DP, Weiland SK, Williams H, The ISAAC Phase Three Study Group (2006) Worldwide time trends in the prevalence of symptoms of asthma, allergic rhinoconjunctivitis, and eczema in childhood: ISAAC phases one and three repeat multicountry cross-sectional surveys. Lancet 368:733–743

Baurecht H, Irvine AD, Novak N, Illig T, Bühler B, Ring J, Wagenpfeil S, Weidinger S (2007) Toward a major risk factor for atopic eczema: meta-analysis of filaggrin polymorphism data. J Allergy Clin Immunol 120:1406–1412

Cavelaars AE, Kunst AE, Geurts JJ, Crialesi R, Grotvedt L, Helmert U, Lahelma E, Lundberg O, Matheson J, Mielck A, Mizrahi A, Rasmussen NK, Regidor E, Spuhler T, Mackenbach JP (1998) Differences in self-reported morbidity by educational level: a comparison of 11 western European countries. J Epidemiol Community Health 52:219–227

Chen JT, Krieger N, Van Den Eeden SK, Quesenberry CP (2002) Different slopes for different folks: socioeconomic and racial/ethnic disparities in asthma and hay fever among 173,859 U.S. men and women. Environ Health Perspect 110:211–216

Cork MJ, Robinson DA, Vasilopoulos Y, Ferguson A, Moustafa M, MacGowan A, Duff GW, Ward SJ, Tazi-Ahnini R (2006) New perspectives on epidermal barrie dysfunction in atopic dermatitis: gene–environment interactions. J Allergy Clin Immunol 118:3–21

D'Amato G, Vitale C, Lanzad M, Molinod A, D'Amato M (2016) Climate change, air pollution, and allergic respiratory diseases: an update. Curr Opin Allergy Clin Immunol 16:434–440

Duggan EM, Sturley J, Fitzgerald AP, Perry IJ, Hourihane JO (2012) The 2002–2007 trends of prevalence of asthma, allergic rhinitis and eczema in Irish schoolchildren. Pediatr Allergy Immunol 23:464–471

Eriksson J, Ekerljung L, Rönmark E, Dahlén B, Ahlstedt S, Dahlén SE, Lundbäck B (2012) Update of prevalence of self-reported allergic rhinitis and chronic nasal symptoms among adults in Sweden. Clin Respir J 6:159–168

Forastiere F, Agabiti N, Corbo GM, Dell'Orco V, Porta D, Pistelli R, Levenstein S, Perucci CA (1997) Socioeconomic status, number of siblings, and respiratory infections in early life as determinants of atopy in children. Epidemiology 8:566–570

Gehring U, Pattenden S, Slachtova H, Antovae T, Braun-Fahrlander C, Fabianova E, Fletcher T, Galassi C, Hoek G, Kuzmin SV (2006) Parental education and children's respiratory and allergic symptoms in the pollution and the young (PATY) study. Eur Respir J 27:95–107

Goh DYT, Chew FT, Quek SC, Lee BW (1996) Prevalence and severity of asthma, rhinitis, and eczema in Singapore school children. Arch Dis Child 74:131–135

Hansen EF, Rappeport Y, Vestbo J, Lange P (2000) Increase in prevalence and severity of asthma in young adults in Copenhagen. Thorax 55:833–836

Holloway JW, Yang IA, Holgate ST (2010) Genetics of allergic disease. J Allergy Clin Immunol 125:81–94

Hsu PS, Campbell DE (2016) A bug's view of allergic airways disease. Paediatr Respir Rev 19:69–74

Hwang JM, Oh SH, Shin MY (2016) The relationships among birth season, sunlight exposure during infancy, and allergic disease Korean. J Pediatr 59:218–225

ISAAC (1991–2012) http://isaac.auckland.ac.nz. Accessed on 11 Jan 2017

Jackson B, Kubzansky LD, Cohen S, Weiss S, Wright RJ (2004) A matter of life and breath: childhood socioeconomic status is related to young adult pulmonary function in the CARDIA study. Int J Epidemiol 33:271–278

Jansson SA, Protudjer JL, Arnlind Heibert M, Bengtsson U, Kallstrom-Bengtsson I, Marklund B, Middelveld RJ, Rentzos G, Sundqvist AC, Akerstrom J, Ostblom E, Dahlen SE, Ahlstedt S (2014) Socioeconomic evaluation of well-characterized allergy to staple foods in adults. Allergy 69:1241–1247

Kim BK, Kim JY, Kang MK, Yang MS, Park HV, Min KU, Cho SH, Kang HR (2016) Allergies are still on the rise? A 6-year nationwide population-based study in Korea. Allergol Int 65:186–191

Lee KS, Rha YH, Oh IH, Choi YS, Choi SH (2016) Socioeconomic and sociodemographic factors related to allergic diseases in Korean adolescents based on the Seventh Korea Youth Risk Behavior Web-based Survey: a cross-sectional study. BMC Pediatr 16:19–25

Lewis SA, Britton JR (1998) Consistent effects of high socioeconomic status and low birth order, and the modifying effect of maternal smoking on the risk of allergic disease during childhood. Respir Med 92:1237–1244

Lewis SA, Weiss ST, Platts-Mills E, Syring M, Gold D (2001) Association of specific allergen sensitization with socioeconomic factors and allergic disease in a population of Boston women. J Allergy Clin Immunol 107:615–623

Liu AH (2002) Endotoxin exposure in allergy and asthma: reconciling a paradox. J Allergy Clin Immunol 109:379–392

Liu AH (2004) Something old, something new: indoor endotoxin, allergens and asthma. Paediatr Respir Rev 5:65–71

Majkowska-Wojciechowska B, Pełka J, Korzon L (2007) Prevalence of allergy, patterns of allergic sensitization and allergy risk factors in rural and urban children. Allergy 62:1044–1050

Marmot MG, Smith GD, Stansfeld S, Patel C, North F, Head J, White I, Brunner E, Feeney A (1991) Health inequalities among British civil servants: the Whitehall II study. Lancet 337:1387–1393

Mercer MJ, Joubert G, Ehrlich RI, Nelson H, Poyser MA, Puterman A, Weinberg EG (2004) Socioeconomic status and prevalence of allergic rhinitis and atopic eczema symptoms in young adolescents. Pediatr Allergy Immunol 15:234–241

Mielck A, Reitmeir P, Wjst M (1996) Severity of childhood asthma by socioeconomic status. Int J Epidemiol 25:388–393

Nathan RA, Meltzer EO, Selner JC, Storms W (1997) Prevalence of allergic rhinitis in the United States. J Allergy Clin Immunol 99:808–816

Okada H, Kuhn C, Feilletand H, Bach JF (2010) The 'hygiene hypothesis' for autoimmune and allergic diseases: an update. Clin Exp Immunol 160:1–9

Pawlińska-Chmara R, Wronka I (2009) Influence of allergic disorders on physical development. Eur J Med Res 14:187–191

Pawlińska-Chmara R, Wronka I, Marchewka J (2013) Effect of socio-economic status on quality of life in people affected with respiratory allergy. Adv Exp Med Biol 788:385–392

Reed CE, Milton DK (2001) Endotoxin-stimulated innate immunity: a contributing factor for asthma. J Allergy Clin Immunol 108:157–166

Samolinski B, Sybilski AJ, Raciborski F, Tomaszewska A, Samel-Kowalik P, Walkiewicz A et al (2009) Prevalence of rhinitis in Polish population according to the ECAP (Epidemiology of Allergic Disorders in Poland) study. Otolaryngol Pol 63:324–330

Samolinski B, Fronczak A, Kuna P, Akdis CA, Anto JM, Bialoszewski AZ, Burney PG, Bush A, Czupryniak A, Dahl R et al (2012) Prevention and control of childhood asthma and allergy in the EU from the public health point of view: an urgent need to fill the gaps. Allergy 67:726–731

Schäfer T, Ruhdorfer S, Weigl L, Wessner D, Heinrich J, Wichmann HE, Ring J (2001) School education and allergic sensitization in adults. Allergy 56:1206–1210

Schmitz R, Atzpodien K, Schlaud M (2012) Prevalence and risk factors of atopic diseases in German children and adolescents – a nationwide health report. Pediatr Allergy Immunol 23:16–723

Sherriff A, Golding J, Alspac Study Team (2002) Hygiene levels in a contemporary population cohort are associated with wheezing and atopic eczema in preschool infants. Arch Dis Child 87:26–29

Strachan DP (1989) Hay fever, hygiene, and household size. BMJ 299:1259–1260

Uphoff E, Cabieses B, Pinart M, Valdés M, Anto JA, Wright J (2015) A systematic review of socioeconomic position in relation to asthma and allergic diseases. Eur Res J 46:364–374

Wang XD, Zheng M, Lou HF, Wang CS, Zhang Y, Bo MY, Ge SQ, Zhang N, Zhang L, Bachert C (2016) An increased prevalence of self-reported allergic rhinitis in major Chinese cities from 2005 to 2011. Allergy 71:1170–1180

Wennergren G, Ekerljung L, Alm B, Eriksson J, Lotvall J, Lundback B (2010) Asthma in late adolescence – farm childhood is protective and the prevalence increase has levelled off. Pediatr Allergy Immunol 21:806–813

Williams HC, Strachan DP, Hay RJ (1994) Childhood eczema: disease of the advantaged? BMJ 308:1132–1135

Advs Exp. Medicine, Biology - Neuroscience and Respiration (2017) 32: 25–31
DOI 10.1007/5584_2017_10

Published online: 3 March 2017

Microbiologic Methods in the Diagnostics of Upper Respiratory Tract Pathogens

J. Kompanikova, A. Zumdick, M. Neuschlova, V. Sadlonova, and E. Novakova

Abstract

Upper respiratory tract infection (URI) is a nonspecific term used to describe acute infections involving the nose, paranasal sinuses, pharynx, and larynx above the vocal cords. The aim of this study was to provide a summary of the most common pathogens of URI and to compare advantages and disadvantages of traditional and new rapid microbiological tests used to identify them. Blood samples were simultaneously examined by the enzyme-linked immunosorbent assay (ELISA) and by the FilmArray Respiratory Panel for eight different pathogens in a total of 15 tests performed in nasopharyngeal swabs. The ELISA method is unable to identify the pathologic agent until the host's immune system elicits a response. The method is readily available in many laboratories at a low cost, which puts less strain on economic resources. The FilmArray® Panel, on the other hand, is more expensive, but it is fast and exact in the identification of a broad spectrum etiologic agents. Nonetheless, since most repiratory tract infections are viral in origin and there is no treatment available, the diagnosis provided by the FilmArray Panel does not provide any additional clinical benefit and thus should be used only whenever necessary on the individual basis.

Keywords

Duagnostics • ELISA method • Microbiological tests • Pathogens • Rapid detection • Respiratory infections • Respiratory panel

1 Introduction

Upper respiratory tract infections (URIs) involve the moist surface of the eyes and eyelids, the nasolacrimal ducts, the middle ear, paranasal sinuses, mastoid air cells, and the main

J. Kompanikova, A. Zumdick, M. Neuschlova, V. Sadlonova, and E. Novakova (✉)
Department of Microbiology and Immunology, Jessenius Faculty of Medicine in Martin, Comenius University in Bratislava, Mala Hora 4B, 036 01 Martin, Slovakia
e-mail: novakova@jfmed.uniba.sk

respiratory passage of the nose and throat as far as the epiglottis and vocal cords. Acute URIs include the common cold, pharyngitis, epiglottitis, and laryngotracheitis (Nester et al. 1995). A variety of viruses, bacteria, fungi, and parasites can infect the respiratory tract. Transmission of organisms occurs by aerosol droplet or direct hand-to-hand contact with infected secretions, with subsequent passage to the nares or eyes. Most URIs are of viral etiology (Dasaraju and Liu 1996). Epiglottitis and laryngotracheitis are exceptions with severe cases likely caused by *Haemophilus influenzae* type B. Bacterial pharyngitis is often caused by *Streptococcus pyogenes*. Bacterial and viral upper respiratory infections produce highly variable clinical symptoms that cannot be used to identify the etiologic agent. Proper treatment depends on the correct identification of a pathogen involved as antibiotics provide little or no benefit with viral infections (Nester et al. 1995).

1.1 Etiology of Upper Respiratory Tract Infections

The URIs are in 69% of viral origin (Mäkelä et al. 1998). Orthomyxoviruses (influenza A and B), paramyxoviruses (parainfluenza and respiratory syncytial viruses), coronaviruses, adenoviruses, and enteroviruses (coxsackie and ECHO viruses) cause common cold. However, most colds are caused by more than 89 types of rhinoviruses (Musher 2003; Cooper et al. 2001). More than 40 strains of adenoviruses cause pharyngitis resembling a strep throat. Rhinoviruses are unresponsive to antibiotics and other medications that control bacterial infections. Antibiotic treatment of adenovirus infections is of no value and sometimes can even be harmful, because it supresses normal bacterial flora and enables the resistant opportunistic pathogens to grow in an uncontrolled way (Nester et al. 1995). Different bacteria, including *Chlamydia pneumoniae, Mycoplasma pneumoniae, Sreptococcus pneumoniae, Bordetella pertussis, Haemophilus influenzae,* and *Staphylococcus aureus* are involved with the upper respiratory system (Murray et al. 2005). The most common URI of bacterial origin is pharyngitis caused by Group A beta-hemolytic streptococci, with *Streptococcous pyogenes* as a main representative, which accounts for 5–10% of pharyngitides (Poole and Portugal 2005). *Streptococcus pneumoniae* and *Haemophilus influenzae* are the most important bacterial pathogens in otitis media and bacterial conjunctivitis. Less commonly, *Mycoplasma pneumoniae, Streptococcous pyogenes,* and *Staphylococcus aureus* are the causative agents in otitis media. A study carried out in the Czech Republic in 2004/05 in 16 different cities among healthy children aged 3–6 years show that the overall carriage of pathogens was 62.8%, with *Streptococcus pneumoniae* 38.1%, *Haemophilus influenzae* 24.9%, *Moraxella catarrhalis* 22.1%, and *Staphylococcus aureus* 16% being the most prevalent (Zemlickova et al. 2006).

1.2 Diagnostic Methods of Upper Respiratory Tract Infections

The diagnosis of URIs is based on a review of symptoms, physical examination, and laboratory tests. Direct identification of bacterial pathogens is based on routine laboratory tests, including growing bacteria in cultures, detection of bacterial metabolic activity, single enzyme tests (catalase, oxidase, urease, or coagulase tests), and molecular methods (Balentine and Siamak 2015; Harvey et al. 2007). Complement fixation test, direct agglutination technique, latex agglutination and enzyme linked immunosorbent assay (ELISA) are traditional serological methods used to detect antibodies in the patient's serum. Traditional diagnostic methods for viral pathogens include growth of the virus in a cell culture, observation of virus particles by electron microscopy, and detection of viral nucleic acid or virus-specific antibodies in the blood (Meneghetti 2016).

1.3 FilmArray Respiratory Panel

The FilmArray Panel is a multiplexed nucleic acid test intended for the simultaneous qualitative detection and identification of multiple respiratory pathogen nucleic acids in nasopharyngeal swabs. This new platform combines automated sample preparation, nucleic acid extraction, and polymerase chain reaction (PCR)-based detection from a single unprocessed sample in 1 h (BioFire Diagnostics; Salt Lake City, UT). This method allows for identification of 21 different respiratory pathogens from a nasopharyngeal swab, 18 of viral etiology and three of bacterial origin (Idaho Technology 2007).

In the present study we seek to determine the individual advantages and disadvantages of different diagnostic tools for pathogens underlying the URIs, as well as under which circumstances a specific method would have an advantage over another one.

2 Methods

2.1 Study Design

The study was approved by the Ethics Commitee of Jessenius Faculty of Medicine in Martin, Slovakia. Results of FilmArray® Respiratory Panel and ELISA tests performed in the Department of Clinical Microbiology of Martin University Hospital were evaluated and compared for the functionalities, advantages, and disadvantages of these methods. We focused on a total number of FilmArray Panel tests, the number of positive results, and the spectrum of pathogens detected in connection with the clinical diagnosis.

2.2 Detection of Pathogens by FilmArray Respiratory Panel

Nasopharyngeal swabs were examined using the FilmArray® Respiratory Panel developed by IDAHO Technology (BioFire Diagnostics; Salt Lake City, UT) for the presence of 21 pathogens (adenovirus, bocavirus, coronavirus HKU1, coronavirus NL63, coronavirus 229E, coronavirus OC43, human metapneumovirus, human rhinovirus/enterovirus, influenza A, influenza A/H1, influenza A/H3, influenza A/H1-2009, influenza B, parainfluenza 1–4, respiratory syncytial virus, *Bordetella pertussis*, *Chlamydophila pneumoniae*, and *Mycoplasma pneumoniae*). To assess the diagnostic efficacy of this new method, a venous blood sample was draw from two patients and it was concurrently examined in the same laboratory for the presence of eight pathogens (adenovirus, influenza A, influenza B, parainfluenza virus 1, respiratory syncytial virus, *Bordetella pertussis*, *Chlamydophila pneumoniae*, and *Mycoplasma pneumoniae*) in 15 tests with the hitherto commonly used ELISA method. Additionally, a hemagglutination inhibition test for influenza A and B was performed in a patient in the Department of Virology of the Regional Institute of Public Health in Banska Bystrica, Slovakia.

3 Results

The FilmArray Panel was performed 15 times to identify URI pathogens in hospitalized patients with acute respiratory infections in January 2016. Positive results were obtained in 8 samples. Four samples turned out positive for human rhinovirus/enterovirus. The other detected pathogens were parainfluenza virus 3, human metapneumovirus, coronavirus OC43, influenza B, and adenovirus. Two different pathogens were identified in one patient (human rhinovirus/enterovirus and adenovirus). Positive results of laboratory tests are summarized in Table 1. These results corresponded with a number of respiratory infections, as diagnosed before hand, such as URIs, pneumonia, acute inflammation of nasopharynx, acute bronchitis, hypothermia unrelated to external temperature, and undefined fever.

For comparison, specimens obtained from two patients were tested simultaneously with

Table 1 Positive results of filmarray respiratory panel – January 2016

Patient	Result	Clinical diagnosis
1	Human rhinovirus/enterovirus	Acute URI
2	Human rhinovirus/enterovirus and adenovirus	Acute URI
3	Human rhinovirus/enterovirus	Acute URI
4	Parainfluenza virus 3	Acute bronchitis
5	Coronavirus OC43	Hypothermia
6	Human metapneumovirus	Pneumonia
7	Human rhinovirus/enterovirus	Undefined fever
8	Influenza B	Acute nasopharyngitis

URI upper respiratory tract infection

Table 2 Results of ELISA tests performed in two patients.

	Patient 1		Patient 2		
Pathogen	Date	Result	Date	Result	Costs
Adenovirus	15.01.2013	Negative	31.01.2013	Negative	IgM = 1.81 €; IgA = 2.07 €
Influenza A	13.01.2013	Negative	31.01.2013	Negative	IgM = 1.91 €; IgA = 1.91 €
Influenza B	13.01.2013	Negative	31.01.2013	Negative	IgM = 1.91 €
Parainfluenza 1	13.01.2013	Negative	31.01.2013	**Positive**	IgM = 2.08 €
Respiratory syncytial virus	12.02.2013	Negative	12.02.2013	Negative	IgM = 1.81 €; IgA = 2.07 €
Bordetella pertussis	13.01.2013	Negative	31.01.2013	Negative	IgM = 1.99 €; IgG = 1.99 €; IgA = 1.99 €
Chlamydophila pneumoniae	18.01.2013	Negative	31.01.2013	Negative	IgG = 1.33 €; IgA = 1.33 €
Mycoplasma pneumoniae	17.01.2013	Negative	31.01.2013	Negative	IgG = 1.33 €; IgA = 1.33 €
Entire testing time	**28 days**		**29 days**		**Total cost – 165 € either patient**

traditional ELISA and FilmArray Panel methods. A venous blood sample and a nasopharyngeal swab were taken from both patients for either method, respectively. In one of these patients, all results were negative and no etiologic agent could be identied by the ELISA method. In total, 28 days were was required to obtain the full panel of results (Table 2; Patient 1). In addition, culture of the specimen obtained from a nasopharyngeal swab from the same patient also was negative for any pathogens. However, nasal swab specimen, examined by the FilmArray Panel, yielded a positive result for human rhinovirus or human enterovirus (Table 3; Patient 1), which was negative when with ELISA.

ELISA performed in a second patient showed an elevation of IgG against parainfluenza virus 1 (IgG positive – 1.53, cut off – 0.709, index – 2.1). This result, however, is inconclusive regarding an acute current infection as it rather confirms a previous encounter with the virus. The result regarding influenza A infection was also negative (IgM – 0.355, cut off – 0.832, index – 0.4; and IgG – 0.083, cut off – 0.381, index – 0.2). The testing time for all eight pathogens amounted to 22 days (Table 2; Patient 2). The examination with the FilmArray Panel in this patient identified the etiologic agent as influenza A/H1-2009 (Table 3; Patient 2). The identification had to be confirmed with a serum hemagglutination inhibition test (HIT), which was done at the Department of Virology of the Regional Institute of Public Health, in Banská Bystrica, Slovakia. The HIT was performed twice 2 weeks apart to assess a possible change in antibody titers. While in the first sample influenza A and B titers were negative, there was a significant rise in the antibody titer against

Table 3 Results of filmarray panel performed in two patients

Pathogen	Date	Patient 1	Patient 2
Adenovirus	15.01.2013	Negative	Negative
Bocavirus	15.01.2013	Negative	Negative
Coronavirus HKU1	15.01.2013	Negative	Negative.
Coronavirus NL63	15.01.2013	Negative	Negative
Coronavirus 229E	15.01.2013	Negative	Negative
Coronavirus OC43	15.01.2013	Negative	Negative
Human metapneumovirus	15.01.2013	Negative	Negative.
Human rhinovirus /enterovirus	15.01.2013	**Positive**	Negative
Influenza A	15.01.2013	Negative	Negative
Influenza A/H1	15.01.2013	Negative	Negative
Influenza A/H3	15.01.2013	Negative	Negative
Influenza A/H1–2009	15.01.2013	Negative	**Positive**
Influenza B	15.01.2013	Negative	Negative
Parainfluenza virus 1, 2, 3, 4	15.01.2013	Negative	Negative
Respiratory syncytial virus	15.01.2013	Negative	Negative
Bordetella pertussis	15.01.2013	Negative	Negative
Chlamydophila pneumoniae	15.01.2013	Negative	Negative
Mycoplasma pneumoniae	15.01.2013	Negative	Negative

Table 4 Hemagglutination inhibition test in Patient 2

Pathogen	First sample 21.01.2013	Second sample 07.02.2013	Costs
Influenza A	Negative	Positive titer – 1:160	Influenza A = 2 € x 2 = 4 €
Influenza B	Negative	Negative titer	Influenza B = 2 € x 2 = 4 €
			Total cost = 8 €

influenza A in the second sample. The HIT assessment took 17 days in all. These results are shown in Table 4.

Health insurance gave 600 points for each of the 15 tests, which makes a total 9000 points, plus 150 points for sample culture from upper respiratory tract and 320 points for a culture form lower respiratory tract. Each insurance point has a value of 0.0066 €. The laboratory therefore received 59.40 € for running the ELISA and additionally 3.10 € for the cultures, which sums up to a total of 62.50 € or 224% of the real material costs. On the other hand, health insurance gave 2500 points for each of the 21 pathogen targets in the FilmArray Panel, which makes a total of 52,500 points and comes to 346.50 € or 210% of the real material costs amounting to 165.00 € per sample. The insurance reimbursed 11.70 € for influenza A and B testing each. Each type of influenza was tested twice giving the cost of 46.80 € in total. The actual laboratory costs for all HIT tests were 8.00 €, which is about 6 times less than the insurance reimbursement.

4 Discussion

Laboratory testing is generally not recommended in the evaluation of upper respiratory infections. Tests for specific pathogens are helpful when therapy depends on the results. Targeted therapy is not available for most viruses that cause URI. Therefore, viral testing is rarely indicated for uncomplicated URIs in the outpatient setting. However, confirmation of a viral condition such as influenza may reduce inappropriate use of antibiotics (Balentine and Siamak 2015). Considering the benfits for patient, the speed to identify the etiologic agent clearly favors the FilmArray System. This method readily identified the etiologic agent in both patients in whom it was applied in the present study, whereas the HIT identified the virus only in the repeat sample of the second patient. However, identification of the etiologic agent did not have any benefit for the first patient with mild symptomatology of coryza because no targeted treatment for human rhinovirus/enterovirus infection was needed. The symptomatology in the second patient was severe and the speed of pathogen identification is crucial for the commencement of appropriate treatment, especially in suspected cases of influenza where early administration of neuraminidase inhibitors significantly reduces mortality rates. In such cases, accuratelly targeted therapy has an enormous benefit for the patient. The laboratory costs to run one examination with different methods showed that the FilmArray multiplex PCR respiratory panel is more expensive than the ELISA, HIT, and the cultivation. One examination with the FilmArray panel brought a 181.50 € *per* patient profit for the laboratory, whereas the profit from running ELISA together with cultivation was 34.62 €, and that from HIT was 38.80 € *per* patient. Therefore, FilmArray respiratory panel is best in terms of profit margin for the laboratory and least favorable for health insurance.

5 Conclusions

Serologic diagnostic methods, such as ELISA and HIT, cannot identify the pathologic agent until the host's immune system elicits a response. However, advantages of those methods are that they are readily available in many laboratories and are least pricey for health insurance. The disadvantage is that the spectrum of pathogens detected is small. A clinical benefit of the FilmArray respiratory panel is that it is quick and exact and may identify a broader spectrum of possible pathogens. Since most URIs are viral in origin and there is no treatment available, the diagnosis provided by the FilmArray panel is not always necessary and the method should be used on an individual basis when clinically justified. From the economic standpoint, FilmArray respiratory panel is the most profitable for the laboratory. In critically ill patients, a spectrum of diagnostic methods should be used to obtain diagnosis as fast as possible. In patients with minor respiratory tract infections, a more rational approach should be undertaken since the speed and accuracy of diagnosis are less crucial. The decision to choose a specific diagnostic method rests with the medical caregiving staff.

Acknowledgments This study was supported by Department of Microbiology and Immunology. We are thankful to our colleagues who provided expertise that greatly assisted the research.

Conflict of Interest The authors declare no conflicts of interest in relation to this article.

References

Balentine JR, Siamak NN (2015) Upper respiratory tract infection. http://www.medicinenet.com/upper_respiratory_infection/page8.htm. Accessed 18 Oct 2016

Cooper RJ, Hoffman JR, Bartlet JF et al (2001) Principles of appropriate antibiotic use for acute pharyngitis in adults: background. Ann Intern Med 134:509–517

Dasaraju PV, Liu C (1996) Infections of the respiratory system. In: Baron S (ed) Medical microbiology, 4th edn Galveston. https://www.ncbi.nlm.nih.gov/books/NBK8142/. Accessed 10 Sept 2016

Harvey RA, Champe PC, Fisher BD (2007) Lippincott's illustrated reviews: microbiology. 2nd edn. Lippincott Williams & Wilkins

Idaho Technology (2007) FilmArray Respiratory Panel (RP) instruction booklet. ftp://ftp.bmgrp.at/Austria/IVD/Idaho%20Manuals%20f%20Helga%20Klinger/RFIT-PRT-0054%20FilmArray%20Respiratory%20Panel%20Instruction%20Booklet/RFIT-PRT-0054%20FilmArray%20Respiratory%20Panel%20Instruction%20Booklet.pdf. Accessed 15 May 2016

Mäkelä MJ, Puhakka T, Ruuskanen O, Leinonen M, Saikku P, Kimpimäki M, Blomqvist S, Hyypiä T, Arstila P (1998) Viruses and bacteria in the etiology of the common cold. J Clin Microbiol 36:539–542

Meneghetti A (2016) Upper respiratory tract infection workup http://emedicine.medscape.com/article/302460-workup. Accessed 24 July 2016

Murray PR, Rosenthal KS, Pfaller MA (2005) Medical microbiology, 5th edn. Elsevier Mosby, Philadelphia

Musher D (2003) How contagious are common respiratory tract infections? N Engl J Med 348:1256–1266
Nester E, Roberts E, Nester M (1995) Microbiology: a human perspective. Wm. C. Brown Publishers, Dubuque
Poole D, Portugal LG (2005) Treatment of rhinosinusitis in the outpatient setting. Am J Med 118:45S–50S
Zemlickova H, Urbaskova P, Adamkova V, Motlova J, Lebedova V, Prochazka B (2006) Characteristics of Streptococcus pneumoniae, *Haemophilus influenzae, Moraxella catarrhalis* and *Staphylococcus aureus* isolated from the nasopharynx of healthy children attending day-care centres in the Czech Republic. Epidemiol Infect 134:1179–1187

Advs Exp. Medicine, Biology - Neuroscience and Respiration (2017) 32: 33–42
DOI 10.1007/5584_2017_19

Published online: 13 April 2017

Seroprevalence of *Bartonella* Species in Patients with Ocular Inflammation

Joanna Brydak-Godowska, Dorota Kopacz, Piotr K. Borkowski, Beata Fiecek, Agata Hevelke, Daniel Rabczenko, Stanisława Tylewska-Wierzbanowska, Dariusz Kęcik, and Tomasz Chmielewski

Abstract

Bartonella species, vector-borne etiologic agents of many systemic or self-limited infections, are responsible for a widening spectrum of diseases in humans, including inflammatory conditions of the eye. The aim of this study was to determine whether there is any relationship between uveitis and the evidence of *Bartonella* spp. infection in the serum, ocular fluid, and cataract mass in patients with intraocular inflammation. Polymerase chain reaction (PCR)-based tests and DNA sequencing were performed on surgery-extracted specimens of intraocular fluid and lens mass of 33 patients. Sera from 51 patients and 101 control subjects were tested for the presence of specific antibodies against *Bartonella* spp. Neither IgM-class antibodies against *Bartonella* spp. nor *Bartonella* spp. DNA were detected. A specific IgG-class antibody was found in 33.3% of the patients with uveitis. The rate of positive *Bartonella* serology was higher among the uveitis patients than that in control subjects. This high rate may in part result from unrecognized indirect

J. Brydak-Godowska, D. Kopacz, and D. Kęcik
Department of Ophthalmology, Medical University of Warsaw, Warsaw, Poland

P.K. Borkowski (✉)
Department of Infectious, Tropical Diseases and Hepatology, Medical University of Warsaw, 37 Wolska Street, 01-201 Warsaw, Poland
e-mail: piotr.k.borkowski@gmail.com

B. Fiecek, S. Tylewska-Wierzbanowska, and T. Chmielewski
Laboratory of Rickettsiae, Chlamydiae and Spirochetes, National Institute of Public Heath-National Institute of Hygiene, Warsaw, Poland

A. Hevelke
Department of Histology and Embryology, Medical University of Warsaw, Warsaw, Poland

D. Rabczenko
Department-Center for Monitoring and Analyses of Population Health Status, National Institute of Public Health-National Institute of Hygiene, Warsaw, Poland

mechanisms rather than the immediate presence and multiplication of *Bartonella* spp. in the eyeball. Nonetheless we believe that screening for *Bartonella* spp. should become part of the diagnostic workup in uveitis.

Keywords
Bartonella spp. • Polymerase chain reaction • Retinitis • Serology • Seroprevalence • Uveitis • Vector-borne pathogens • Zoonosis

1 Introduction

Bartonellosis is a vector-borne zoonosis caused by various species of the aerobic Gram-negative bacteria of the genus *Bartonella*. To date some 25 species of *Bartonella* have been described and three of them are known to cause uveitis (Fiecek et al. 2012, Drancourt et al. 2008). Pets (dogs, cats) and wild animals, depending on the geographical region rabbits, moles, squirrels, deer or coyotes are reservoirs of infection while fleas, flies, lice, ticks, and other blood-sucking arthropods are the vectors. *Bartonella* spp. are animal pathogens and man is an accidental host. Geographical distribution of different *Bartonella* species vary, with *B. henselae* serologically detected in 3.7–54.6% of cats depending on the region. In Poland, infections are mostly caused by *B. henselae* and *B. quintana*, with specific antibodies present in 48% of cats in the capital city of Warsaw region to 86% in other regions. Infection is transmitted by a bite or scratch by an infected cat or by contact of damaged skin or mucous membranes with the feces of the cat flea (*Ctenocephalides felis*) (Reis et al. 2011; Podsiadły et al. 2009; Breitschwerdt and Kordick 2000).

Up to 7.5% of ticks in Austria and 4.8% in Poland are infected by *B. henselae* (Müller et al. 2016; Sytykiewicz et al. 2012). Exposure to tick bites is a factor substantially contributing to the preponderance of a seropositive response to *B. henselae* in the forestry workers and farmers in Poland (Zając et al. 2015). *Bartonella* penetrates the skin and infects the host cell by sticking to epithelial cells, including vascular endothelial cells, using trimeric autotransporter adhesins (TAAs) secreted from the bacterial outer membrane (Szczesny et al. 2008). The contagion enters the host cell cytoplasm by phagocytosis, where it multiplies in phagosome vesicles. *B. henselae* can also use a transport-dependent reorganization of the host cell cytoskeleton, with the utilization of the Rho family of GTPases: Cdc42, Rho, and Rac (Truttmann et al. 2011).

Typical signs of *B. henselae* infection include papules or blisters at the bite or scratch site, enlargement of regional lymph nodes, fever, headache, and enlargement of the spleen. In severe cases, typically in immunocompromised patients, neurological disorders, parenchymatous inflammation, hepatitis, osteomyelitis, encephalitis, glomerulonephritis, endocarditis, or pneumonia may develop (Mączka and Tylewska-Wierzbanowska 2012; Zenone 2011; Podsiadły et al. 2009). Occasionally, *Bartonella* infection can mimic malignancy (Mazur-Melewska et al. 2015). Signs and symptoms of systemic disease may be accompanied by a variety of ocular manifestations, which are observed in 5–10% of patients. The most commonly reported ocular manifestations in patients with cat scratch disease are Parinaud oculoglandular syndrome (POGS) and optic neuritis (5% of patients), and retinitis (2% of patients) (Cunningham and Koehler 2000; Ormerod and Dailey 1999; Carithers 1985). A review of the literature shows that infection with *B. henselae* can cause inflammation in various locations in the eyeball: iritis, vitreous inflammation, posterior uveitis, choroiditis, retinitis and retinal vasculitis (Kalogeropoulos et al. 2011; Terrada et al. 2009; Drancourt et al. 2008; Goldstein et al. 2001).

This study seeks to define the prevalence of *Bartonella* spp. infection in ocular uveitis and retinitis. We addressed the issue by detecting specific IgM- and IgG-class antibodies in the serum and in the intraoperatively extracted specimens of ocular fluid and cataract mass in patients. This work is a continuation of previous studies carried out in patients with cataract, but without associated uveitis (Chmielewski et al. 2014).

2 Methods

The study was approved by the Bioethics Committee of the National Institute of Public Health-National Institute of Hygiene in Warsaw, Poland. The ocular inflammation group consisted of 51 patients with chronic uveitis or retinitis (F/M – 36/15) of the mean age of 50.0 ± 14.4 years (range 21–71 years). The disease duration was 8 months to 10 years and some of the patients suffered from chronic comorbidities (sarcoidosis and arthritis in four patients each and multiple sclerosis in two patients). The diagnosis was set according to the Standardization of Uveitis Nomenclature Study Group (Jabs et al. 2005). Laboratory and imaging (fluorescein angiography, optical coherence tomography, ultrasonography, and plain chest X-rays) examinations were performed in all patients to determine the cause of uveitis or retinitis. Of the 51 patients, 33 underwent cataract phacoemulsification surgery and glaucoma surgery (trabeculectomy) was performed in three patients during follow-up.

Blood samples from patients were collected for serological testing. The level of specific antibodies against *Bartonella* spp. were evaluated with the indirect immunofluorescence method using *Bartonella* IFA IgG test (Focus Diagnostics; Cypress, CA), where the *B. henselae* and *B. quintana* cultured in the Vero cell line were used as the diagnostic antigens. The titers of ≥64 were considered positive. The control group consisted of 101 blood donors (F/M – 47/54) of the mean age of 48.5 ± 10.4 years (range 31–65 years).

Specimens of intraocular fluid and the lens mass removed from the eye during surgery were tested by polymerase chain reaction (PCR) for *Bartonella* DNA. Bacterial DNA from the collected material was extracted with the QIAamp Tissue kit (QIAGEN GmbH; Hilden, Germany) according to the manufacturer's recommendations. Extracted DNA was amplified to detect fragments complementary to NADH dehydrogenase subunit gamma (nuoG) gene of *Bartonella* spp. (amplified PCR product of 346 bp). In two patients, owners of cats, extra material was collected from three cats, consisting of two samples of cat saliva and one blood sample, to check for *Bartonella* spp. DNA. Data are reported as means ± SD. Statistical elaboration was done with the chi-squared test. A p-value <0.05 defines statistically significant differences.

3 Results

The ocular inflammation group included 20 patients with anterior uveitis, 21 patients with posterior uveitis, nine patients with panuveitis, and one patient with intermediate uveitis. In 28 patients, uveitis was unilateral and in 23 it was bilateral. Idiopathic uveitis was diagnosed in 28 patients and Fuchs heterochromic uveitis in 12 patients. Specific *Bartonella* spp. antibodies were detected in 17 (33.3%) serum samples from patients (F/M – 5/12) aged 25–74 years.

Positive *Bartonella* serology was found in six (40.0%) patients of the 15 patients who did not undergo surgery and in eleven (30.5%) of the 36 patients who underwent eye-surgery. Of those with positive *Bartonella* serology, IgG-class antibody titer was 64 in nine (17.6%) patients, 128 in six (11.7%) patients, and 512 and 1024 in one (2%) patient each. IgM-class antibodies were not detected. Among the patients with positive *Bartonella* spp. serology, nine had cataract phacoemulsification and two had glaucoma surgery (trabeculectomy) performed. The results for individual patients are presented in Table 1 and summary of patients characteristics in Table 2.

DNA of *Bartonella* spp. was not detected in the material collected during surgery from the anterior eye chamber and lens mass. Nor was it detected in the three cats tested either. In the

Table 1 Demographic, clinical, and serological results of individual patients

Patient nr	Age (year)	Gender	Comorbidity/ systemic manifestations	Serologic testing-IgG class antibody titer	Ocular findings	Ocular fluid examination	Cat owner	Animal-inflicted scratch or bite	Treatment
1	44	F	No	256	Active posterior veitis/ perivasculitis	No	No	No	Topical steroid and NSAID; systemic doxycycline/prednisolone
2	54	F	No	1024	Active panuveitis/ vasculitis	Phacoemulsification	Yes	Yes (cat)	Topical steroid and NSAID; systemic doxycycline prednisolone
3	52	F	A history of interferon treatment for HCV	128	Inactive retinitis	Phacoemulsification	No	No	Treated in the past with acyclovir/ valacyclovir/prednisolone
4	41	F	Persistent upper respiratory infection	128	Active choroiditis	No	No	No	Treated in the past with systemic doxycycline/azithromycin
5	25	F	Upper respiratory infection	64	Active retinochoroiditis	No	No	No	Systemic doxycycline/prednisone
6	43	M	No	64	Inactive posterior uveitis/ vasculitis	No	No	No	No
7	58	F	Sarcoidosis	64	Active panuveitis/ vasculitis	Phacoemulsification	No	No	Topical steroid and NSAID; systemic prednisone
8	29	F	Upper respiratory infection, CSD	512	Fuchs	Phacoemulsification	Yes	Yes (cat)	Systemic doxycycline
9	50	F	No	64	Fuchs	Trabeculectomy	No	No	Topical steroid and NSAID
10	61	F	Diabetes	64	Posterior uveitis/ perivasculitis	No	No	No	Topical steroid and NSAID/ prednisolone
11	62	F	Hypertension	128	Inactive posterior uveitis/ vasculitis	Phacoemulsification	No	No	No
12	56	F	No	64	Fuchs	Phacoemulsification	No	No	Topical steroid and NSAID

13	59	F	No	128	Fuchs	Trabeculectomy	No	Yes (monkey)	Topical steroid and NSAID
14	48	F	No	64	Inactive posterior uveitis/ vasculitis	Phacoemulsification	No	No	No
15	74	F	No	128	Anterior uveitis	Phacoemulsification	No	No	Topical steroid and NSAID
16	28	M	No	64	Reccurent retinochoroiditis	No	No	No	Systemic doxycycline/prednisolone
17	38	F	No	64	Active panuveitis	Phacoemulsification	No	No	Topical steroid and NSAID; systemic doxycycline/prednisolone/ acyclovir/valacyclovir

HCV hepatitic C virus, *NSAID* nonsteroidal anti-inflammatory drugs, *CSD* cat scratch disease

Table 2 Patients and results

	Patients (n; %)
Gender	
F	15 (88.2)
M	2 (11.8)
Serologic testing (titer)	
64	9 (52.9)
128	5 (29.4)
256	1 (5.9)
512	1 (5.9)
1024	1 (5.9)
Co-morbidities	
No	10 (58.8)
Yes	7 (41.2)
Cat owner	
No	15 (88.2)
Yes	2 (11.8)
Scratches	
No	14 (82.4)
Yes	3 (17.6)

Table 3 Positive serology findings by surgery group

	n	Positive serology (n; %)	p
No surgery	15	6 (40.0)	0.51
Surgery	36	11 (30.6)	

control group, specific IgG-class antibodies to *B. henselae* antigen were detected in one person in titer 128. Thus, positive serology was observed in 17/51 (33.3%) patients with chronic uveitis and in 1/101 (1.0%) person in the control group; the difference between the two groups was significant ($p < 0.001$).

A higher percentage of patients with positive serology was observed in a group of patients without surgery than in those surgically treated (40.0% vs. 30.6%), although the difference between these two groups failed to reach statistical significance (Table 3).

4 Discussion

Bartonelloses is a group of zoonotic infectious diseases occurring worldwide. Clinical diagnosis may be difficult to establish because infection manifestations may widely vary from mild flu-like symptoms to life-threatening conditions such as pneumonia, myocarditis, perinephric abscess, or encephalitis (Rising et al. 2016; Atıcı et al. 2014; Rondet et al. 2012; Kaiser et al. 2011; Boulouis et al. 2005; Eskow et al. 2001). Some patients diagnosed with cat scratch disease do not develop fever or lymphadenopathy (Robert et al. 2012; Chomel et al. 2006; Tsuneoka and Tsukahara 2006; Kawasaki and Wilson 2003; Carithers 1985). The disease course depends on the immune status of the host (Lee et al. 2015; Psarros et al. 2012; Resto-Ruiz et al. 2003).

In spite of a considerable body of research into *Bartonella* spp. as an infectious agent in humans, relatively little is known about human humoral and cellular immune responses triggered by the exposure to the pathogen. Studies in rodent models demonstrate asymptomatic bacteremia for 5–6 days after infection followed by the formation of bacterial aggregates and the invasion of mature erythrocytes with the intraerythrocytic bacteremia usually subsiding after 8–10 weeks. At that time, the immune response occurs, both humoral (increases in IgM and then IgG levels) and cellular (increased T-cell counts). Studies of classical facultative intracellular parasites, such as *Listeria monocytogenes* and *Mycobacterium tuberculosis*, have demonstrated a major role of the cellular response in defending the host against intracellular pathogens. The cellular response, however, is effective only against those microbes that parasitize the cells capable of presenting the major histocompatibility complex (MHC) class I-dependent pathways of antigen presentation. The MHC class I molecules occur on the surface of all nucleated cells, but erythrocytes do not have nuclei and hence do not express MHC class I molecules, although there may be some expression of MHC class II molecules on the surface of erythrocytes. Thus, proteinic *Bartonella* antigens are protected against sensitized T-lymphocytes with a greater role of the humoral response manifested as increased titers of specific IgG immunoglobulins capable of destroying 'free' *Bartonella* forms before their entry into erythrocytes (Harms and Dehio 2012; Koesling et al. 2001).

Most published studies demonstrate the association of *B. henselae* infection and various eye disorders, including optic neuritis, uveitis, or choroidopathy while the link with *B. quintana* or *B. grahami* is less commonly reported (Goldstein et al. 2001; Kalogeropoulos et al. 2011; Drancourt et al. 2008). Diagnosis of bartonellosis or cat scratch disease is based on a history of scratch or bite by a pet cat or a wild animal or insect (tick, flea, or louse) bite, clinical manifestations, and serological tests. According to Carithers (1985), *B. henselae* infection most commonly affects the lymphatic system, followed by the eye. There are reports of patients with ocular lesions and specific *B. henselae* antibodies, but in some cases without any systemic manifestations, who deny any contact with animals (Manousaridis et al. 2015; Robert et al. 2012; Terrada et al. 2009; Kawasaki and Wilson 2003; Solley et al. 1999).

Of the 51 patients with different types of uveitis observed in the present study, 17 (33.3%) had antibodies against *Bartonella* spp. and two were the cat owners for several years and had been frequently bitten or scratched by their pets. According to Blanco Ramos et al. (1998) approximately 28.9% of cat owners have positive *Bartonella* serology but no systemic manifestations. In the present study, patient number 8 reported poorly healing scratches on his hands and arms and low grade fever, cat scratch disease was diagnosed and treatment with doxycycline was instituted for symptomatic relief. The remaining patients had sporadic contact with animals of their neighbors or friends but did not give a history of animal-inflicted skin injuries. None of the patients gave a history of tick-bite, although most could be unaware of their potential exposure to ticks and tick-borne infections possibly contracted during leisure time. Patient number 13 reported that she had been bitten by a monkey 20 years before and the wound healed well after topical treatment. According to Breitschwerdt and Kordick (2000), bacteremia is observed in 50–90% of animal populations, which may be the cause of frequent infections transmitted to humans.

Tick bites are underestimated by both specialists and the general public as a potential source of infections caused by *Bartonella* spp. It has been documented that from 1.7 to 4.8% of ticks in Poland are infected with *B. henselae*. Further studies are needed to establish the rate of transmission from ticks to humans and whether it depends on the length of time the tick remains attached to the host's skin. Based on the data from a Polish study of Sytykiewicz et al. (2012) and assuming that the rate of transmission may be similar to that of *Borrelia*, the number of people infected with *Bartonella* spp. from ticks could equal a third of borreliosis cases. In a study of Terrada et al. (2009) every third patient with ocular manifestations reported contact with animals. In patients with full-blown cat scratch disease, ocular manifestations usually occur within 1 to 3–8 weeks after onset of systemic disease, most commonly flu-like symptoms (Cunningham and Koehler 2000; Wade et al. 2000; Ormerod and Dailey 1999).

Patients in the present study suffered from chronic ocular inflammation. In none of them were IgM-class antibodies detected by ELISA, which might be related to the fact that all had eye disorders for a relatively long time. IgM-class antibodies appear immediately after infection and persist for up to 3 months (Carithers 1985). The first ocular manifestations of *Bartonella* infection are seen late, usually approximately 8 weeks after infection and IgM-class antibodies are seldom detected (Fiecek et al. 2012). On the other hand, IgG-class antibodies were detected in 17 patients (33.3%) and their titers varied from 64 to 1024. Positive *Bartonella* serology was found in 6 (40.0%) of the 15 patients who did not undergo surgery and in 11 (30.5%) eye-surgery patients. According to Sander et al. (1998), IgG titers of ≥256 in patients with systemic manifestations of cat scratch disease indicate an active disease, as noted in two patients. Low IgG-class antibody titers of 64–256 may indicate the beginning of a disease or much earlier exposure to the pathogen, as noted in 15 patients (Wade et al. 2000). Specific IgG-class antibodies against *Bartonella* spp. may be detected up to 2 years after infection. In the present study, serological tests detected *Bartonella* spp. antibodies more frequently in patients with uveitis than in controls, although the two groups differed in size.

Of the 17 uveitis patients with positive *Bartonella* serology, anterior uveitis was diagnosed in 5 patients, including four patients with Fuchs heterochromic uveitis. The latter is a chronic anterior uveitis with documented viral or protozoal etiology (Jad et al. 2013; Babu and Murthy 2012; Kongyai et al. 2012; Liu et al. 2011). Posterior uveitis was diagnosed in the remaining 12 patients and in seven patients it was associated with retinal periphlebitis. Vascular changes described in patients with bartonellosis are associated with vascular endothelial damage (Eiger-Moscovich et al. 2016; Manousaridis et al. 2015; Robert et al. 2012; Dehio 2005). In 11 of the 17 patients with positive *Bartonella* serology, nine procedures of phacoemulsification cataract extraction and two glaucoma surgery procedures (trabeculectomy) were performed. No *Bartonella* spp. DNA was found in the intraoperative specimens removed during cataract extraction or trabeculectomy (anterior chamber fluid and phacoemulsified lens material). We have previously performed a similar study in patients with cataract without coexisting uveititis. A total of 109 specimens of anterior chamber fluid and phacoemulsified lens material from the nucleus and cortex have been examined and *Bartonella* spp. DNA has been found in specimens from 2 patients. One of them had a history of central retinal vein thrombosis and the other was diagnosed with common variable immunodeficiency (Chmielewski et al. 2014).

The absence of *Bartonella* spp. DNA may be due to the long-term inflammation affecting the eye treated with the topical or systemic treatment. Of the 17 patients with long-term uveitis in the present study, only one, a cat owner, was found to have a full-blown cat scratch disease. Interestingly, the remaining patients had no evident systemic symptoms of bartonellosis. It cannot be excluded, however, that they had had at the beginning a pauci-symptomatic infection or their symptoms were diagnosed as a generalized seasonal infection which quickly resolved. In a study of Drancourt et al. (2008), among 1321 patients with uveitis of unknown etiology, which included unilateral and bilateral, granulomatous and non-granulomatous inflammation and panuvetitis, *Bartonella* spp. was found in 21 patients, which was confirmed by the examination of ocular fluid in three patients and by serological testing in 18 patients. *B. henselae* was identified in nine, *B. quintana* in seven, and *B. grahami* in four patients. In three other patients, uveitis was due to undetermined *Bartonella* species. In all cases of uveitis caused by *B. henselae,* the Huston genotype determined by DNA sequencing was the offending microorganism. In the present study, we did not observe any ocular manifestations typical of *B. henselae* infection, such as neuroretinitis with a macular star. A limitation of this study is a small population of patients with positive *Bartonella* spp. serology, which was borderline in a number of cases and the fact that ocular fluid samples were investigated in eye-surgery patients only.

5 Conclusions

We found a higher proportion of positive *Bartonella* serology among uveitis patients than in control subjects, which was also higher relative to the general Polish population. This high prevalence of positive *Bartonella* serology is likely to result from currently unrecognized indirectly acting mechanisms, e.g., some degree of immune deficiency, than from the presence and multiplication of *Bartonella* spp. in the eyeball, because its DNA was not identified in the eyeball. We believe that screening for *Bartonella* spp. infection should be included in the diagnostic workup of uveitis, even when there is no conclusive history of infection and there are no systemic manifestations.

Acknowledgments The authors would like to acknowledge the contribution of NIPH-NIH in Warsaw for support of serology tests and PCR.

Conflicts of Interest The authors declare no conflicts of interest in relation to this article.

References

Atıcı S, Kadayıfcı EK, Karaaslan A, Toper MH, Celikel CA, Soysal A, Bakır M (2014) Atypical presentation of cat-scratch disease in an immunocompetent child with serological and pathological evidence. Case Rep Pediatr 2014:397–437. doi:10.1155/2014/397437

Babu K, Murthy GJ (2012) Chikungunya virus iridocyclitis in Fuchs' heterochromic iridocyclitis. Indian J Ophthalmol 60(1):73–74

Blanco Ramos JR, Oteo Revuelta JA, Martínez de Artola V, Ramalle Gómara E, García Pineda A, Ibarra Cucalón V (1998) Seroepidemiology of *Bartonella henselae* infection in a risk group. Rev Clin Esp 198 (12):805–809

Boulouis HJ, Chang CC, Henn JB, Kasten RW, Chomel BB (2005) Factors associated with rapid emergence of zoonotic *Bartonella* infections. Vet Res 36 (3):383–410

Breitschwerdt EB, Kordick DL (2000) *Bartonella* infection in animals: carriership, reservoir potential, pathogenicity, and zoonotic potential for human infection. Clin Microbiol Rev 13(3):428–438

Carithers HA (1985) Cat scratch disease: an overview based on a study of 1,200 patients. Am J Dis Child 139:1124–1133

Chmielewski T, Brydak-Godowska J, Fiecek B, Rorot U, Sędrowicz E, Werenowska M, Kopacz D, Hevelke A, Michniewicz M, Kęcik D, Tylewska-Wierzbanowska S (2014) Bacterial tick-borne diseases caused by *Bartonella* spp., *Borrelia burgdorferi* sensu lato, *Coxiella burnetii*, and *Rickettsia* spp. among patients with cataract surgery. Med Sci Monit 20:927–931

Chomel BB, Boulouis HJ, Maruyama S, Breitschwerdt EB (2006) *Bartonella* spp. in pets and effect on human health. Emerg Infect Dis 12(3):389–394

Cunningham ET Jr, Koehler JE (2000) Ocular bartonellosis. Am J Ophthalmol 130:340–349

Dehio C (2005) Bartonella-host-cell interactions and vascular tumour formation. Nat Rev Microbiol 3 (8):621–631

Drancourt M, Berger P, Terrada C, Bodaghi B, Conrath J, Raoult D, LeHoang P (2008) High prevalence of fastidious bacteria in 1520 cases of uveitis of unknown etiology. Medicine (Baltimore) 87(3):167–176

Eiger-Moscovich M, Amer R, Oray M, Tabbara KF, Tugal-Tutkun I, Kramer M (2016) Retinal artery occlusion due to *Bartonella henselae* infection: a case series. Acta Ophthalmol 94(5):367–370

Eskow E, Rao RV, Mordechai E (2001) Concurrent infection of the central nervous system by *Borrelia burgdorferi* and *Bartonella henselae*: evidence for a novel tick-borne disease complex. Arch Neurol 58 (9):1357–1363

Fiecek B, Chmielewski T, Tylewska-Wierzbanowska S (2012) *Bartonella* spp. infections with particular emphasis on eye diseases. Post Mikrobiol 51 (1):47–53. (Article in Polish)

Goldstein DA, Mouritsen L, Friedlander S, Tessler HH, Deepak PE (2001) Acute endogenous endophthalmitis due to *Bartonella henselae*. Clin Infect Dis 33:718–721

Harms A, Dehio C (2012) Intruders below the radar: molecular pathogenesis of *Bartonella* spp. Clin Microbiol Rev 25(1):42–78

Jabs DA, Nussenblatt RB, Rosenbaum JT (2005) Standardization of uveitis nomenclature for reporting clinical data. Results of the first international workshop. Am J Ophthalmol 140(3):509–516

Jad A, Céline T, Bahram B, Phuc L, Nathalie C (2013) Fuchs' heterochromic cyclitis: a post-infectious manifestation of ocular toxoplasmosis? Int Ophthalmol 33 (2):189–194

Kaiser PO, Riess T, O'Rourke F, Linke D, Kempf VA (2011) *Bartonella* spp.: throwing light on uncommon human infections. Int J Med Microbiol 301(1):7–15

Kalogeropoulos C, Koumpoulis I, Mentis A, Pappa C, Zafeiropoulos P, Aspiotis M (2011) *Bartonella* and intraocular inflammation: a series of cases and review of literature. Clin Ophthalmol 5:817–829

Kawasaki A, Wilson DL (2003) Mass lesions of the posterior segment associated with *Bartonella henselae*. Br J Ophthalmol 87(2):248–249

Koesling J, Aebischer T, Falch C, Schülein R, Dehio C (2001) Cutting edge: antibody-mediated cessation of hemotropic infection by the intraerythrocytic mouse pathogen *Bartonella grahamii*. J Immunol 167 (1):11–14

Kongyai N, Sirirungsi W, Pathanapitoon K, Tananuvat N, Kunavisarut P, Leechanachai P, de Groot-Mijnes JD, Rothova A (2012) Viral causes of unexplained anterior uveitis in Thailand. Eye 26(4):529–534

Lee RA, Ray M, Kasuga DT, Kumar V, Witherspoon CD, Baddley JW (2015) Ocular bartonellosis in transplant recipients: two case reports and review of the literature. Transpl Infect Dis 17(5):723–727

Liu Y, Takusagawa HL, Chen TC, Pasquale LR (2011) Fuchs heterochromic iridocyclitis and the rubella virus. Int Ophthalmol Clin 51(4):1–12

Mączka I, Tylewska-Wierzbanowska S (2012) Heart disease as a late complications of zoonosis transmitted by a ticks. Post Mikrobiol 51(1):37–45. (Article in Polish)

Manousaridis K, Peter S, Mennel S (2015) Cat-scratch-disease-associated macular oedema treated with intravitreal ranibizumab. Acta Ophthalmol 93 (2):168–170

Mazur-Melewska K, Jończyk-Potoczna K, Mania A, Kemnitz P, Szydłowski J, Służewski W, Figlerowicz M (2015) The significance of *Bartonella henselae* bacteria for oncological diagnosis in children. Infect Agent Cancer 10:30. doi:10.1186/s13027-015-0025-x

Müller A, Reiter M, Schötta AM, Stockinger H, Stanek G (2016) Detection of *Bartonella* spp. in *Ixodes ricinus* ticks and *Bartonella* seroprevalence in human populations. Ticks Tick Borne Dis 7(5):763–767

Ormerod LD, Dailey JP (1999) Ocular manifestations of cat-scratch disease. Curr Opin Ophthalmol 10:209–216

Podsiadły E, Sapiejka E, Dąbrowska-Bień J, Majkowski J, Tylewska-Wierzbanowska S (2009) Diagnostics of cat scratch disease and present methods of bartonellosis recognition – a case report. Pol Merkur Lekarski 26 (152):131–135. Article in Polish

Psarros G, Riddell J 4th, Gandhi T, Kauffman CA, Cinti SK (2012) *Bartonella henselae* infections in solid organ transplant recipients: report of 5 cases and review of the literature. Medicine (Baltimore) 91 (2):111–121

Reis C, Cote M, Le Rhun D, Lecuelle B, Levin ML, Vayssier-Taussat M, Bonnet SI (2011) Vector competence of the tick *Ixodes ricinus* for transmission of *Bartonella birtlesii*. PLoS Negl Trop Dis 5(5):e1186. doi:10.1371/journal.pntd.0001186

Resto-Ruiz S, Burgess A, Anderson BE (2003) The role of the host immune response in pathogenesis of *Bartonella henselae*. DNA Cell Biol 22(6):431–440

Rising T, Fulton N, Vasavada P (2016) Splenorenal manifestations of *Bartonella henselae* infection in a pediatric patient. Case Rep Radiol 2016:7803832. doi:10.1155/2016/7803832

Robert MC, Saab M, Rezende FA (2012) Presumed choroidal neovascularization secondary to *Bartonella henselae*. Can J Ophthalmol 47(4):e9–e10. doi:10.1016/j.jcjo.2012.03.010

Rondet B, Sarret C, Lacombe P, Rouveyrol F, Chenel C, Romaszko JP, Labbé A (2012) Neurological symptoms with *Bartonella henselae* infection: report on 2 pediatric cases. Arch Pediatr 19(8):823–826

Sander A, Posselt M, Oberle K, Bredt W (1998) Seroprevalence of antibodies to *Bartonella henselae* in patients with cat scratch disease and in healthy controls: evaluation and comparison of two commercial serological tests. Clin Diagn Lab Immunol 5(4):486–490

Solley WA, Martin DF, Newman NJ, King R, Callanan DG, Zacchei T, Wallace RT, Parks DJ, Bridges W, Sternberg P Jr (1999) Cat scratch disease: posterior segment manifestations. Ophthalmology 106 (8):1546–1553

Sytykiewicz H, Karbowiak G, Werszko J, Czerniewicz P, Sprawka I, Mitrus J (2012) Molecular screening for *Bartonella henselae* and *Borrelia burgdorferi* sensu lato co-existence within *Ixodes ricinus* populations in central and eastern parts of Poland. Ann Agric Environ Med 19(3):451–456

Szczesny P, Linke D, Ursinus A, Bär K, Schwarz H, Riess TM, Kempf VA, Lupas AN, Martin J, Zeth K (2008) Structure of the head of the *Bartonella* adhesin BadA. PLoS Pathog 4(8):e1000119. doi:10.1371/journal.ppat.1000119

Terrada C, Bodaghi B, Conrath J, Raoult D, Drancourt M (2009) Uveitis: an emerging clinical form of *Bartonella* infection. Clin Microbiol Infect 15(Suppl 2):132–133

Truttmann MC, Misselwitz B, Huser S, Hardt WD, Critchley DR, Dehio C (2011) *Bartonella henselae* engages inside-out and outside-in signaling by integrin β1 and talin1 during invasome-mediated bacterial uptake. J Cell Sci 124:3591–3602

Tsuneoka H, Tsukahara M (2006) Analysis of data in 30 patients with cat scratch disease without lymphadenopathy. J Infect Chemother 12(4):224–226

Wade NK, Levi L, Jones MR, Bhisitkul R, Fine L, Cunningham ET Jr (2000) Optic disk edema associated with peripapillary serous retinal detachment: an early sign of systemic *Bartonella henselae* infection. Am J Ophthalmol 130(3):327–334

Zając V, Wójcik-Fatla A, Dutkiewicz J, Szymańska J (2015) *Bartonella henselae* in eastern Poland: the relationship between tick infection rates and the serological response of individuals occupationally exposed to tick bites. J Vector Ecol 40(1):75–82

Zenone T (2011) Systemic *Bartonella henselae* infection in immunocompetent adult presenting as fever of unknown origin. Case Rep Med 2011:183937. doi:10.1155/2011/183937

Advs Exp. Medicine, Biology - Neuroscience and Respiration (2017) 32: 43–51
DOI 10.1007/5584_2017_15

Published online: 3 March 2017

Body Composition, Anthropometric Indices and Hydration Status of Obstructive Sleep Apnea Patients: Can Cachexia Coexist with Obesity?

Barbara Kuźnar-Kamińska, Marcin Grabicki, Tomasz Trafas, Monika Szulińska, Szczepan Cofta, Tomasz Piorunek, Beata Brajer-Luftmann, Agata Nowicka, Barbara Bromińska, and Halina Batura-Gabryel

Abstract

The aim of this study was to elucidate body composition, anthropometric indices, and hydration status in obstructive sleep apnea (OSA) patients, taking into account different disease stages, gender, and the possibility of the presence of cachexia. There were 98 OSA patients and 23 control subjects enrolled into the study. All study participants underwent polysomnography examination. Body mass index (BMI), fat mass index (FMI), fat free mass, muscle mass, body cell mass, total body water, and extracellular and intracellular water were evaluated. The neck, abdominal, and waist circumference was measured. We found that overweight and obesity were present in 96% of patients. Cachexia was present in one OSA individual with comorbidities. Apnea-hypopnea index correlated with the neck and waist circumference, and with BMI in OSA patients. All muscle indices and water contents above outlined were significantly higher in severe OSA compared with control subjects. BMI, FMI, neck circumference, and extracellular water were greater in a subset of severe OSA compared with a moderate OSA stage. The female OSA patients had a higher FMI than that present in males at a comparable BMI. We conclude that the most body composition indices differed significantly between severe OSA patients and control subjects. A higher FMI in females at a

B. Kuźnar-Kamińska (✉), M. Grabicki, T. Trafas, S. Cofta, T. Piorunek, B. Brajer-Luftmann, A. Nowicka, and H. Batura-Gabryel
Department of Pulmonology, Allergology and Respiratory Oncology, Poznan University of Medical Sciences, 84 Szamarzewskiego Street, 60-569 Poznan, Poland
e-mail: kaminska@ump.edu.pl

M. Szulińska
Department of Internal Diseases, Metabolic Disorders and Hypertension, Poznan University of Medical Sciences, Poznan, Poland

B. Bromińska
Department of Endocrinology, Metabolism and Internal Medicine, Poznan University of Medical Sciences, Poznan, Poland

comparable BMI could be due to a discordance between BMI and FMI. Cachexia occurs rarely in OSA and seems to coexist with comorbidities.

Keywords
Apnea-hypopnea index • Body composition • Body water content • Cachexia • Hydration • Nutritional status • Obesity • Obstructive sleep apnea

1 Introduction

Obstructive sleep apnea (OSA) is a disorder characterized by repeated respiratory events during sleep caused by obstruction of upper airways. The occurrence of sleep-disordered breathing leads to the destruction of sleep architecture and desaturation events (De 2013). The incidence of OSA is 2% in women and 4% in man. In the obese population, the percentage of individuals with positive diagnosis increases to over 20% (Young et al. 2002). There are a number of risk factors for OSA, the most important are gender (males are more predisposed), obesity, age between 45 and 65, smoking, consumption of alcohol, sleeping pills, and nasal occlusion (Plywaczewski et al. 2013; Al Lawati et al. 2009; Young et al. 2004).

Nutritional status is an integral part of the assessment of patients with OSA. It is well known that OSA occurs mainly in obese patients; however, it is also diagnosed in slim individuals. Nutritional status of OSA patients could be linked to the inflammatory process (Tasali and Ip 2008). The visceral fat mass in obese persons secretes inflammatory agents, but irrespective of obesity intermittent hypoxia can stimulate inflammatory pathways which through other mediators regulate appetite and body mass (Bonsignore et al. 2012; Yin et al. 2009). In some chronic inflammatory diseases, e.g., chronic obstructive pulmonary disease, heart failure, or chronic kidney disease, inflammation is connected with body composition changes and can cause cachexia (von Haehling and Anker 2010; von Haehling et al. 2007). Cachexia is a specific form of malnutrition, which is characteristic of chronic inflammatory diseases with a decrease in lean body mass while the fat mass remains unchanged, which makes it relatively higher (Evans et al. 2008). Cachectic patients can suffer from malnutrition, be normally nourished, or even be obese. Considering the above facts, measurement of body composition is necessary in a complex assessment of diseases with inflammation.

The human body is a model built of two compartments: fat mass (FM) and fat-free mass (FFM). The FM constitutes about 25% of body mass and is primarily created from adipose tissue. The main component of FFM is water, which comprises about 73% and is a component of muscle mass (MM). The FFM weight can be divided into extracellular (ECW) and intracellular (ICW) weight. The former is a replaceable part of the energy while the latter are fixed elements such as tendons, plasma, or interstitial fluid. Bioelectrical impedance analysis (BIA) is a reliable, non-invasive, safe, and effective technique of measuring body composition. The BIA is based on the phenomenon of electrical resistance and the difference in the electrical conductivity of human body tissues.

There have been several studies presenting a detailed assessment of nutritional status in OSA patients, but the occurrence of cachexia in OSA patients has not yet been examined. In the present study, therefore, we set out to examine the body composition, anthropometric indices and the hydration status in OSA patients, taking into account the disease severity and patient gender.

Further, the study seeks to determine whether cachexia could be an accompaniment of OSA.

2 Methods

The study protocol was approved by the Institutional Review Board at Poznan University of Medical Sciences in Poznan, Poland (permit no. 641/15) and all patients gave written informed consent for the proposed procedures. The protocol conforms to the principles of the Helsinki Declaration of the World Medical Association.

2.1 Patients

One hundred and twenty one adult patients (96 males and 25 females; mean age 57.8 ± 15.2 and 61.9 ± 12.9, respectively) seen in a pulmonary outpatient clinic with symptoms pointing to the possibility of OSA were enrolled into the study. The symptoms consisted of overnight awakenings, dyspnea during the night, snoring, overnight arousals, nocturia, a feeling of suffocation or choking, daytime sleepiness, morning headaches, fatigue, impaired intellectual function, and problems with memory and concentration. The patients were further referred to the Pulmonary Department for polysomnography (PSG) and other diagnostic tests. The presence of OSA was confirmed in 98 patients who became the study group. A control group consisted of 23 persons, described as 'simple snorers' in whom OSA was excluded. The OSA group was stratified into three subsets of patients based on disease severity: 47 persons with severe, 36 with moderate, and 15 with mild OSA.

A detailed medical history was taken from all patients on admission. Co-existing diseases were verified based on the medical documentation. In the OSA group, 7% of patients were treated for heart insufficiency, 4% were diagnosed with chronic obstructive pulmonary disease (COPD), and 7% had chronic renal diseases.

2.2 Polysomnography

Diagnostic tests were conducted in a sleep laboratory during hospitalization in the Department of Pulmonology, Allergology and Respiratory Oncology of Poznan University of Medical Sciences in Poland. All study participants underwent complete overnight, supervised polysomnography using an Embla S4000 sleep diagnostics setup (Natus Medical Inc., Pleasanton, CA). The following variables were recorded: electroencephalogram (EEG), electromyogram (EMG), electrooculogram (EOG), electrocardiogram (ECG), blood oxygen saturation (pulse-oximetry), nasal and mouth airflows (thermistor, nasal cannula), breathing movements of the thorax and abdomen, sounds while breathing (snoring), and sleep position.

Severity of OSA was determined according to the American Academy of Sleep Medicine (AASM) guidelines, with the apnea/hypopnea index (AHI) ≥ 5 adopted as the diagnostic criterion. OSA severity was determined as follows: mild – $5 \leq \text{AHI} < 15$ (15 patients; F/M – 4/11), moderate – $15 \leq \text{AHI} < 30$ (36 patients; F/M – 7/29), and severe – $\text{AHI} \geq 30$ (47 patients; F/M – 7/40). The Epworth Sleepiness Scale (ESS) was used to measure daytime sleepiness.

2.3 Body Composition

Body composition was evaluated by using a bioimpedance method (AKERN BIA 101; SMT Medical, Wuerzburg, Germany). The test was done in the morning following the polysomnographic examination after fasting without beverages and physical activity for at least 12 h before the measurement, and in a room at controlled temperature of 21 °C. Weight and height of each patient was taken using a beam scale with the accuracy of 0.1 kg and 1 cm, respectively.

This non-invasive test involves the placement of two electrodes on the person's right hand and right foot. Tissues that contain large amounts of fluid and electrolytes, such as the blood, have high conductivity; and fat and bone slow the signal down. The flow of current is affected by the amount of water in the body. The device measures how the signal is impeded through different types of tissue. The following indices of body composition and hydration status were determined: body mass (BM), fat mass (FM), fat free mass (FFM), muscle mass (MM), body cell mass (BCM), extracelullar water (ECW), intracellular water (ICW), total body water (TWC). The indexes of these parameters, BMI, FMI, FFMI, MMI, BCMI, ECWI, and ICWI, were calculated as component of body mass/height in m^2.

The BMI normal range was adopted as 18.5–24.9 kg/m^2. Overnutrition was diagnosed with $25.0 \leq$ BMI ≤ 29.9 kg/m^2. Obesity was defined as BMI ≥ 30 kg/m^2 and malnutrition as BMI < 18.5 kg/m^2 (White et al. 2012). Cachexia was defined as FFMI < 16.0 kg/m^2 for males and 15.0 kg/m^2 for females (Vanltallie et al. 1990).

Neck circumference (NC), abdominal circumference (AC), and waist circumference (WC) were measured using a non-stretchable plastic tape with the subjects standing upright.

2.4 Statistical Analysis

Data were expressed as means ± SD. Differences between groups of patients were evaluated using the Student's *t*-test or Mann-Whitney U test as required. Associations were examined using Pearson's or Spearman's rank correlation coefficients, depending on data distribution. A p-value < 0.05 defined statistically significant changes. The evaluation was performed using a commercial GraphPad Prism ver. 5.0 packet.

3 Results

3.1 Body Composition and Hydration in OSA Patients and Control Subjects

Patient characteristics are presented in Table 1. The BMI in OSA patients ranged from 22.0 to 55.4 kg/m^2, with the mean of 34.1 ± 7.0 kg/m^2. Ninety four (77.7%) patients were overweight, 23 (19.0%) obese, and 4 (3.4%) had the BMI within normal range. None of the patients had underweight.

The FFMI in OSA patients ranged from 7.7 to 31.6 kg/m^2, with the mean of 21.8 ± 3.9 kg/m^2. The lowest FFMI of 7.7 kg/m^2 was present in an overweight patient with BMI 27.2 kg/m^2. This patient suffered from COPD and chronic renal failure. Other than that, FFMI was normal, but in 13 (13.2%) patients it was in the lower range of values ($\leq$18 kg/m^2). Among those patients, two suffered from chronic renal failure, and one from heart insufficiency.

Table 1 Body composition and water content characteristics of OSA patients and control subjects

	OSA	Control	p-value
Age (year)	58.6 ± 14.8	54.0 ± 11.8	0.059
Height (m)	1.72 ± 0.09	1.73 ± 0.09	ns
Weight (kg)	100.7 ± 21.1	83.2 ± 14.5	<0.001
BMI (kg/m^2)	34.1 ± 7.0	27.8 ± 4.0	<0.001
AHI (events/h)	37.3 ± 23.8	<5.0	<0.001
NC (cm)	43.2 ± 3.4	39.7 ± 3.2	<0.001
AC (cm)	112.6 ± 15.5	98.9 ± 12.7	<0.001
WC (cm)	107.8 ± 13.7	95.6 ± 11.1	<0.001
FMI (kg/m^2)	12.1 ± 5.0	7.6 ± 3.8	<0.001
FFMI (kg/m^2)	21.8 ± 3.9	20.2 ± 3.2	ns
BCMI (kg/m^2)	12.4 ± 3.2	12.4 ± 2.3	ns
MMI (kg/m^2)	15.2 ± 2.7	14.7 ± 3.5	ns
TWCI (L/m^2)	16.7 ± 2.6	15.0 ± 2.2	0.005
ECWI (L/m^2)	6.1 ± 0.9	7.3 ± 1.5	<0.001
ICWI (L/m^2)	9.1 ± 2.1	9.4 ± 1.6	ns

BMI body mass index, *NC* neck circumference, *AC* abdominal circumference, *WC* waist circumference, *FMI* fat mass index, *FFMI* fat free mass index, *BCMI* body cell mass index, *MMI* muscle mass index, *TWCI* total water content index, *ECWI* extracelullar water index, *ICWI* intracellular water index, *ns* non-significant

Body composition of OSA patients differed significantly from that of the control subjects with regard to fat body components. The weight, BMI, and FMI of OSA patients were higher than those of 'simple snorers'. However, indices describing lean body mass, FFMI and MMI, did not differ between the two groups (Table 1).

OSA patients had a larger neck, abdominal, and waist circumferences than those in control subjects. These circumferences were related to AHI in OSA patients (Fig. 1a, b and c). The AHI also correlated with BMI, FMI, and FFMI (Fig. 1d, e and f). The mean TBWI and ECWI were significantly higher in OSA patients than those in control subjects. This relationship was not confirmed for ICWI.

3.2 Body Composition and Hydration with Reference to OSA Severity

A comparison of body composition, anthropometric indices, and water content between patients of different OSA severity and control subjects is presented in Figs. 2 and 3. There was no difference among any of the indices between the mild and moderate OSA groups. There were, however, significant differences in all indices, except for MMI and ICWI, between severe OSA and control subjects. In addition, indices related to fat mass such as BMI, FMI, and NC differed significantly between severe and moderate OSA groups. NC also differed between severe and mild OSA. TBWI differed between severe OSA and control subjects ($p < 0.05$).

3.3 Body Composition and Hydration with Reference to Gender

A comparison of body composition, anthropometric indices, and water content in male and female OSA patients is presented in detail in Table 2. Both men and women were of comparable age, and had a comparable AHI and daytime sleepiness assessed with the ESS scale. On average, men's weight was significantly greater than that of women, at a comparable BMI. All women (BMI range from 25.2 to 48.9 kg/m^2) and 95% of men (BMI range from 25.7 to 55.4 kg/m^2) were overweight or obese. The remaining 5% minority of men had BMI within normal range. Body cell

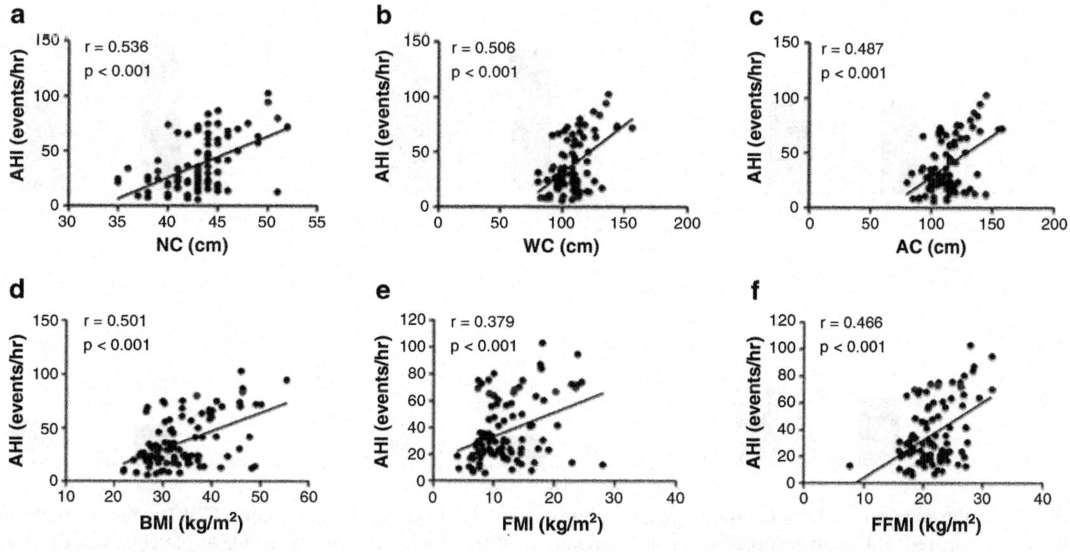

Fig. 1 Correlation between apnea-hypopnea index (AHI) and (**a**) neck circumference (NC), (**b**) waist circumference (WC), (**c**) abdomen circumference (AC), (**d**) body mass index (BMI), (**e**) fat mass index (FMI), and (**f**) free mass index (FFMI) in OSA patients

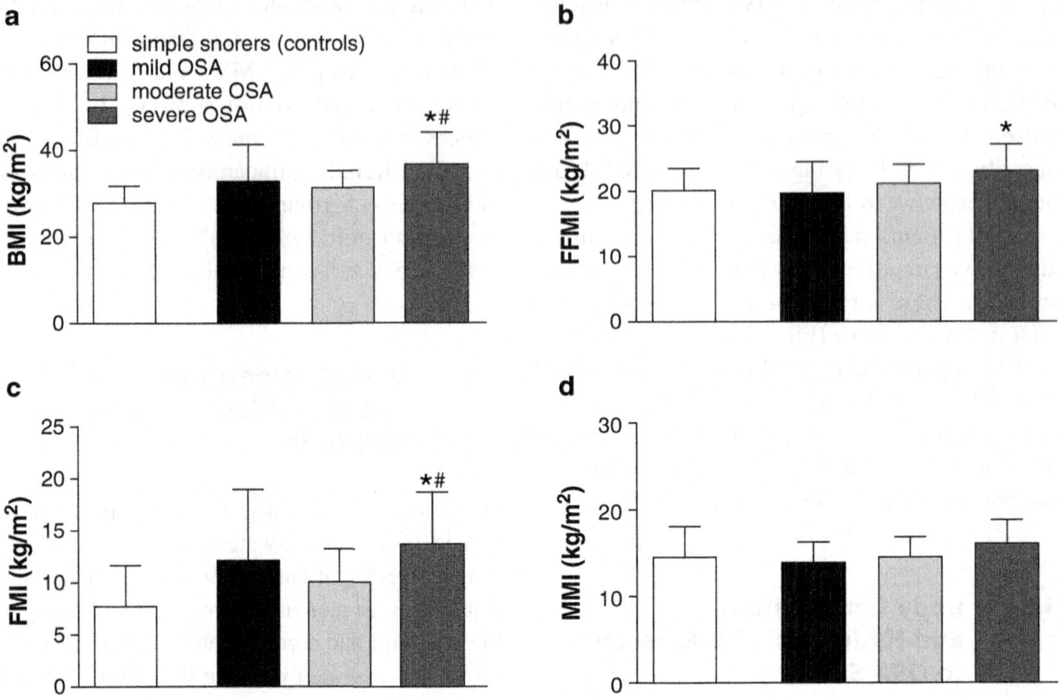

Fig. 2 Body composition in relation to the severity of OSA compared with control subjects: (**a**) body mass index (BMI); (**b**) fat free mass index (FFMI); (**c**) fat mass index (FMI); and (**d**) muscle mass index (MMI); $^*p < 0.05$ for differences between severe OSA patients and controls; $\#p < 0.05$ for differences between severe and moderate OSA

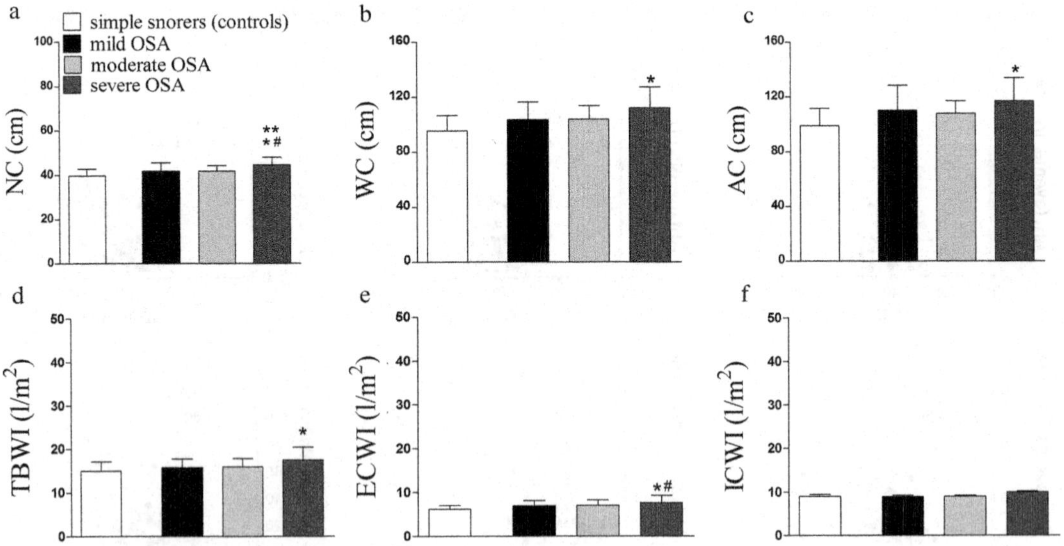

Fig. 3 Anthropometric indices in relation to the severity of OSA compared with control subjects: (**a**) neck circumference (NC), (**b**) waist circumference (WC), (**c**) abdomen circumference (AC), (**d**) total body water indicator (TBWI), (**e**) extracellular water indicator (ECWI), and (**f**) intracellular water indicator (ICWI); $^*p < 0.05$ for differences between severe OSA patients and controls; $\#p < 0.05$ for differences between severe and moderate OSA; $^{**}p < 0.05$ for differences between severe and mild OSA

Table 2 Gender-dependent differences in body composition and water content in OSA patients

	OSA men	OSA women	p-value
Height (m)	1.75 ± 0.08	1.60 ± 0.07	<0.001
Weight (kg)	103.2 ± 21.2	90.1 ± 17.5	0.017
BMI (kg/m^2)	33.8 ± 6.8	35.4 ± 7.7	ns
AHI (events/h)	38.8 ± 24.4	30.6 ± 20.2	ns
ESS	9.8 ± 5.0	8.3 ± 5.5	ns
NC (cm)	44.0 ± 2.8	39.2 ± 2.9	<0.001
AC (cm)	113.9 ± 14.6	106.5 ± 18.2	ns
WC (cm)	109.2 ± 12.4	101.5 ± 17.7	0.034
FMI (kg/m^2)	11.2 ± 4.3	16.2 ± 5.9	<0.001
FFMI (kg/m^2)	22.4 ± 2.6	19.2 ± 2.7	0.001
BCMI (kg/m^2)	12.7 ± 2.4	11.0 ± 1.2	0.004
MMI (kg/m^2)	15.6 ± 2.8	13.5 ± 1.4	0.002
TWCI (L/m^2)	17.1 ± 2.6	14.8 ± 1.9	0.004
ECWI (L/m2)	7.5 ± 1.4	6.3 ± 1.2	<0.001
ICWI (L/m^2)	9.6 ± 1.7	8.5 ± 0.9	<0.001

BMI body mass index, *AHI* apnea/hyponea index, *ESS* Epworth sleepiness scale, *NC* neck circumference, *AC* abdominal circumference, *WC* waist circumference, *FMI* fat mass index, *FFMI* fat free mass index, *BCMI* body cell mass index, *MMI* muscle mass index, *TWCI* total water content index, *ECWI* extracelullar water index, *ICWI* intracellular water index, *ns* non-significant

mass, muscle mass and fat free mass were significantly greater in men. Likewise, water content measured in all compartments was greater in men than in women.

4 Discussion

In the present study, we sought to determine body composition and water content distribution in OSA patients with a different intensity of disease symptoms. We also took into account nutritional status of patients and possible gender-dependent differences in disease presentation. We found that a number of indices inherent for obesity and metabolic syndrome, such as body mass index, neck, abdomen, and waist circumferences, muscle mass indicators were significantly enhanced in OSA patients compared with the control group consisting of 'simple snorers'. These indices were also associated with disease severity. The results demonstrate that AHI was most related to cervical obesity. These results are different than those of Lovin et al. (2010) who have reported that abdominal adiposity was the best predictor of OSA severity. Previous studies have confirmed that obesity in OSA is related to inflammation. Acute sleep deprivation activates the hypothalamic–pituitary–adrenal axis, inhibits the growth hormone (GH)/insulin-like growth factor-1 (IGF-1) axis, increases proinflammatory cytokines, induces insulin resistance, all of which facilitates the development of metabolic syndrome (Kyrou et al. 2006; McEwen 2006; Spiegel et al. 2004). On the other hand, chronic inflammation could be connected with cachexia, a feature of other chronic diseases (von Haehling et al. 2007). In COPD patients, obesity and co-existing cachexia have been demonstrated (Furutate et al. 2011). Taking that into account, some OSA patients could be potentially predisposed to cachexia. The meaning of cachexia in OSA is elusive. In the present study, we found only one OSA patient with decreased FFMI independently of other fat mass indices, who was a patient with COPD and chronic kidney disease. Thirteen (13.2%) OSA patients had lean body mass indicators within the limit of normal values. Among them, three patients had co-existing chronic diseases that could be related to cachexia. Moreover, we show that FFMI and MMI did not differ between OSA patients and control subjects, although BMI and FMI were significantly higher in OSA patients. The predominance of obesity over cachexia could be a consequence of enhanced appetite and positive energy balance, both of which are connected with release of specific inflammatory factors in response to chronic sympatho-adrenal activation in OSA (Nadeem et al. 2013; Spiegel et al. 2004).

Apart from the little explained role of energy balance in OSA, the role of body water content has been studied less often. We found an increased index of total water content, while the extracellular water was decreased in OSA patients. This result is different from increased extracellular and decreased intracellular water content reported in a study of Kosacka et al. (2013). Changes in body water distribution in OSA patients could be a consequence of increased atrial natriuretic peptide excretion and disturbance of the renin-angiotensin-aldosterone activity (O'Hearn et al. 2009; Redolfi et al. 2009)

In the present study we demonstrate that BMI, FMI, and neck circumference, i.e., indices closely associated with obesity, were significantly greater in severe than moderate OSA. No appreciable differences were observed in body composition and anthropometric indices between mild and moderate OSA. The lack of differences of indices between mild and moderate OSA subsets could stem from a relatively small number of patients classified as having mild OSA.

Gender comparison demonstrates that there were differences in body composition, anthropometric indices, and hydration status in OSA patients despite the matching regarding BMI and AHI. Women had a greater fat mass with a lower content of fat free mass and muscle mass, which is in line with the results present in the general population. Previous studies have shown that a greater fat mass at a comparable BMI in females could be due to a discordance between fat mass and BMI present in women but not in men, which can, in turn, result in a reduced bone mineral density in women (Zhu et al. 2017).

Taken together, the present study confirmed the relationship between OSA and obesity. Most of body composition indices differed between severe OSA and control subjects, but there were no major differences between OSA severity subsets. There were also gender differences, notably characterized by BMI and muscle fat discordance in women suffering from OSA. Cachexia rarely occurs in OSA patients and when it does it seems a feature of accompanying chronic comorbidities such as COPD, renal insufficiency, or heart failure. The possible presence of cachexia in OSA should be further explored in another study designs.

Conflicts of Interest The authors declare no conflicts of interest in relation to this article.

References

Al Lawati NM, Patel SR, Ayas NT (2009) Epidemiology, risk factors, and consequences of obstructive sleep apnea and short sleep duration. Prog Cardiovasc Dis 51:285–293

Bonsignore MR, Esquinas C, Barceló A, Sanchez-de-la-Torre M, Paternó A, Duran-Cantolla J, Marín JM, Barbé F (2012) Metabolic syndrome, insulin resistance and sleepiness in real-life obstructive sleep apnoea. Eur Respir J 39:1136–1143

De BW (2013) Obstructive sleep apnea/hypopnea syndrome. Panminerva Med 55:191–195

Evans WJ, Morley JE, Argiles J et al (2008) Cachexia: a new definition. Clin Nutr 27:793–799

Furutate R, Ishii T, Wakabayashi R, Motegi T, Yamada K, Gemma A, Kida K (2011) Excessive visceral fat accumulation in advanced chronic obstructive pulmonary disease. Int J Chron Obstruct Pulmon Dis 6:423–430

Kosacka M, Korzeniewska A, Jankowska R (2013) The evaluation of body composition, adiponectin, C-reactive protein and cholesterol levels in patients with obstructive sleep apnea syndrome. Adv Clin Exp Med 22:817–824

Kyrou I, Chrousos GP, Tsigos C (2006) Stress, visceral obesity, and metabolic complications. Ann N Y Acad Sci 1083:77–110

Lovin S, Bercea R, Cojocaru C, Rusu G, Mihăescu T (2010) Body composition in obstructive sleep apneahypopnea syndrome bio-impedance reflects the severity of sleep apnea. Multidiscip Respir Med 5:44–49

McEwen BS (2006) Sleep deprivation as a neurobiologic and physiologic stressor: allostasis and allostatic load. Metabolism 55:S20–S23

Nadeem R, Molnar J, Madbouly EM, Nida M, Aggarwal S, Sajid H, Naseem J, Loomba R (2013) Serum inflammatory markers in obstructive sleep apnea: a meta-analysis. J Clin Sleep Med 9:1003–1012

O'Hearn DJ, Gold AR, Gold MS, Diggs P, Scharf SM (2009) Lower extremity edema and pulmonary hypertension in morbidly obese patients with obstructive sleep apnea. Sleep Breath 13:25–34

Plywaczewski R, Brzecka A, Bielicki P, Czajkowska-Malinowska M, Cofta S, Jonczak L, Radliński J, Tażbirek M, Wasilewska J, Polish Society of Lung Diseases (2013) Sleep related breathing disorders in adults – recommendations of Polish Society of Lung Diseases. Pneumonol Alergol Pol 81:221–258. (Article in Polish)

Redolfi S, Yumino D, Ruttanaumpawan P, Yau B, Su MC, Lam J, Bradley TD (2009) Relationship between overnight rostral fluid shift and obstructive sleep apnea in nonobese men. Am J Respir Crit Care Med 179:241–246

Spiegel K, Tasali E, Penev P, Van CE (2004) Brief communication: sleep curtailment in healthy young men is associated with decreased leptin levels, elevated ghrelin levels, and increased hunger and appetite. Ann Intern Med 141:846–850

Tasali E, Ip MS (2008) Obstructive sleep apnea and metabolic syndrome: alterations in glucose metabolism and inflammation. Proc Am Thorac Soc 5:207–217

Vanltallie TB, Yang MU, Heymsfield SB, Funk RC, Boileau RA (1990) Height-normalized indices of the body's fat-free mass and fat mass: potentially useful indicators of nutritional status. Am J Clin Nutr 52:953–959

von Haehling S, Anker SD (2010) Cachexia as a major underestimated and unmet medical need: facts and numbers. J Cachex Sarcopenia Muscle 1:1–5

von Haehling S, Doehner W, Anker SD (2007) Nutrition, metabolism, and the complex pathophysiology of cachexia in chronic heart failure. Cardiovasc Res 73:298–309

White JV, Guenter P, Jensen G, Malone A, Schofield M, Academy of Nutrition and Dietetics Malnutrition Work Group, A.S.P.E.N. Malnutrition Task Force, A.S.P.E.N. Board of Directors (2012) Consensus statement of the Academy of Nutrition and Dietetics/ American Society for Parenteral and Enteral Nutrition: characteristics recommended for the identification and documentation of adult malnutrition (undernutrition). J Acad Nutr Diet 112:730–738

Yin J, Gao Z, He Q, Zhou D, Guo Z, Ye J (2009) Role of hypoxia in obesity-induced disorders of glucose and lipid metabolism in adipose tissue. Am J Physiol Endocrinol Metab 296:E333–E342

Young T, Peppard PE, Gottlieb DJ (2002) Epidemiology of obstructive sleep apnea: a population health perspective. Am J Respir Crit Care Med 165:1217–1239

Young T, Skatrud J, Peppard PE (2004) Risk factors for obstructive sleep apnea in adults. JAMA 291:2013–2016

Zhu K, Hunter M, James A, Lim EM, Cooke BR, Walsh JP (2017) Discordance between fat mass index and body mass index is associated with reduced bone mineral density in women but not in men: the Busselton healthy ageing study. Osteoporos Int 28:259–268

Advs Exp. Medicine, Biology - Neuroscience and Respiration (2017) 32: 53–68
DOI 10.1007/5584_2017_18

Published online: 6 April 2017

Frailty and Primary Sarcopenia: A Review

Eli Carmeli

Abstract

Increasing longevity, coupled with rising frailty and sarcopenia of aging, significantly affects function and quality of life of older adults. This review discusses the definition, assessment, and management of frailty and sarcopenia, and examines the relationship between them. Medline, Scopus and Psychoinfo databases were searched using the keywords frailty, sarcopenia, aging, and functional disability. The findings are that frailty and sarcopenia are often assessed clinically with such methods such as DeXA, CT scan, MRI, bioelectrical impedance, or anthropometry. Frailty and sarcopenia differentially affect older adults. Both conditions are characterized by decreased energy reserves and resistance to external and internal stressors, resulting in susceptibility to fatigue, comorbidity, sedentary life style, functional decline, hospitalization, quality of life, and even death. The estimated prevalence of frailty with sarcopenia is relatively low; however, the condition requires early detection and careful management.

Keywords

Aging • Frailty • Functional disability • Lifespan • Longevity • Sarcopenia

1 Introduction

This article reviews the body of knowledge related to frailty and sarcopenia. According to a recent report by the WHO, the proportion of people aged over 65 increases faster than that of any other age group (WHO 2015). Today, life expectancy of a 65-year old person is 3–4 years more than it was 20 years ago; meaning more people are getting old, and more old people are getting older. Aging is accompanied by gradual, yet progressive changes in all biological systems, and it affects physical, cognitive, psychological, and social abilities. More specifically, the aging process is accompanied by changes in physical activity, decline in lean body mass, and reduced

E. Carmeli (✉)
Department of Physical Therapy, University of Haifa, 199 Aba Khoushy Ave, Haifa 3498838, Israel
e-mail: ecarmeli@univ.haifa.ac.il

muscle mass and strength, leading to primary sarcopenia (Cruz-Jentoft et al. 2014; Landi et al. 2014).

Primary sarcopenia is an age-related syndrome mainly represented by a reduction in muscle mass and strength and difficulties performing daily living activities. Frailty is a physiological decline in many biological systems resulting in poor health, weight loss, moderate to severe dependency in daily living activities, recurrent hospitalizations, and death (Turner et al. 2014; Rodríguez-Mañas et al. 2013; Heuberger 2011; Sternberg et al. 2011; Xue 2011; Lang et al. 2009; Walston et al. 2006; Fried et al. 2001; Hamerman 1999). The link between frailty and sarcopenia is not well-described in the literature. The conditions are interrelated conditions and lead to difficulties performing daily living activities and increased disability (Peters et al. 2012; Raîche et al. 2008; Ravaglia et al. 2008; Saliba et al. 2001). A review by McCleary et al. (2014) has stressed that older adults with frailty or sarcopenia are prone to complications during cancer treatment. Guidelines published in 2010 by the American College of Surgeons indicate the importance of assessing both frailty and sarcopenia prior to oncologic surgery in the elderly (Chow et al. 2012).

Despite the challenges in clarifying the definition of these conditions, assessing them is even more complex and has less consensus. Some studies evaluated the presence of frailty or sarcopenia by using functional assessments, such as the Katz Index (Figueiredo et al. 2013), the Short Physical Performance Battery (SPPB) (Chang et al. 2014), and the Timed Up and Go (TUG) test (Hassani et al. 2015). Furthermore, it has been shown that function is the most important prognostic measure in predicting co-morbidities and mortality (Inouye et al. 1998). It is also reasonable to assume the presence of a cognitive decline related to frailty with or without primary sarcopenia (Bowen 2012; Vermeulen et al. 2011). Therefore, assessment of the older adult should performed be according to The International Classification of Functioning, Disability and Health (ICF) (Azzopardi et al. 2016; Cao and Morley 2016). This includes several psychosocial, cognitive, functional, recreational, and participation domains. Early screening for frailty may be valuable for preventing or minimizing sarcopenia and hopefully reversing it. The prevalence of these individual and combined medical diagnoses is of interest because of their association with health prevention, prognosis, and intervention. The purpose of this review is to describe frailty and sarcopenia, define their assessment and management, and to examine the relationship between them.

2 Sarcopenia Definition

The term sarcopenia was first defined by Irwin Rosenberg in 1989 (Rosenberg 1989). The currently accepted definition includes three phases of the health condition, as introduced by the European Working Group on Sarcopenia in Older People (EWGSOP). This group has developed a practical clinical definition and diagnostic criteria for age-related sarcopenia (Cruz-Jentoft et al. 2014). The first phase is defined as pre-sarcopenia, where a reduction in muscle mass is observed. This is followed by decreased muscle strength (i.e., sarcopenia), followed by the third phase, severe sarcopenia, that occurs when, in addition to reduction in muscle mass and strength, physical performance declines. Another definition of sarcopenia has been introduced in Rome, Italy in November 2009 by the International Working Group on Sarcopenia. That definition emphasizes the evaluation of walking speed and measuring muscle mass (Fielding et al. 2011). Recently, the term muscle quality (MQ) has been suggested in the clinical setting to describe the ratio between muscle power and muscle mass (Barbat-Artigas et al. 2012). Thus, MQ can be regarded as a clinical marker of muscle efficiency. Whatever the definition, the prevalence of sarcopenia is reported as up to one-third of older adults (Walston et al. 2006; Fried et al. 2001).

3 Anthropometric Assessment

The most frequent methods used in clinical practice are anthropometric. These are noninvasive, quantitative techniques for determining body composition by analyzing specific dimensions, such as height and weight; skin-fold thickness; and waist, hip, and chest circumference. Body mass index (BMI) is one of the most popular and important indices to assess body composition (Kulkarni et al. 2013). References for cut-off points for low or high BMI for males and females were established in diverse populations (Yu et al. 2015; Araujo et al. 2010). Interestingly, the cut-off points for Caucasians are higher than for Asians (Silva et al. 2010).

Anthropometrics are simple, clinical tools that can be easily used for sarcopenia because they are the most portable, commonly applicable, inexpensive, and non-invasive techniques for assessing the size, proportion, and composition of the human body. However, their validity is limited when applied to individuals due to a large prediction errors and because the cut-off points to identify low muscle mass still need to be defined. Therefore, if a patient is identified as being at-risk for sarcopenia based on anthropometrics, additional measurements of muscle mass with dual-energy X-ray absorptiometry (DeXA) is recommended (Villani et al. 2013, 2014).

3.1 Leg and Arm Circumference

Limb circumferences correlate with appendicular muscle mass and reflect both health and nutritional status. They also predict physical performance, health and survival in older people (Landi et al. 2010, 2014; Rolland et al. 2003). The validity of measuring lower limb circumference to detect sarcopenia is still debatable and the inter-tester reliability is weak.

3.2 Skin Fold Thickness

Loss of skin elasticity and subcutaneous fat can provide a general idea of body composition. The association between obesity and sarcopenia is widely reported (Visser et al. 2002). It is accepted that decreased muscle mass and/or muscle strength occurs with adipose cell infiltration into muscle or through lipogenesis of satellite cells (adult stem cells) that differentiate to adipocytes instead of myocytes (Vettor et al. 2009). A caliper is used to measure subcutaneous tissue. The skin fold pinch can be taken at many sites around the body. The most common are the triceps, a site 1 cm below the inferior angle of the scapula, and another site immediately above the iliac crest (top of hip bone) on the most lateral aspect. Because of high errors involved, it is usually not appropriate to convert skin fold measures to percentage body fat (%BF). Therefore, it is best to use the sum of the three sites to monitor and compare body fat measures. Interestingly, a study by Tamura et al. (2012) has suggested evaluating sarcopenia of the lingual muscles by measuring tongue thickness. The idea behind this is that elderly people often suffer from malnutrition caused by dysphagia. This often leads to sarcopenia, which may compromise oral function. Thus, tongue thickness could be related to nutritional status. However, it must be measured using ultrasonography.

4 Muscle Mass Assessment

Decreased muscle mass is defined as a slow, progressive reduction in the cross-sectional area CSA) of muscle fibers (Wei et al. 2016). Initially, the sarcoplasm volume decreases, then the fast twitch (type II) muscle fibers get shorter, followed by a shrinking diameter of slow twitch (type I) muscle fibers (Drey 2011). Corresponding to this reduction in fiber CSA, is a progressive increase in dense interstitial collagen and adipose connective tissue (Scott et al. 2015). At a later stage, termed the ultra-structural level, sarcopenic fibers show evidence of disorganization, including misalignment of Z bands, dissociation of the T system and a displacement of sarcosomes due to distraction of intermediate filaments that hold the sarcosome in place. These changes are similar to those seen in muscular dystrophy (Malatesta et al. 2014). Structural changes also occur in the extracellular matrix, including an increase in collagen concentration and a change in the elastic fiber system. These changes in the elastic fiber result in increased tissue stiffness, which also plays a role in the overall decrease in muscle function.

When measuring muscle mass, the cross-sectional area, the interstitial connective tissue, and the subcutaneous adipose tissue should all be considered. In the absence of a gold standard, a few general and specific outcome measures are used to assess muscle mass. The general, indirect tools to assess muscle mass include biochemical markers, nutrition intake, body mass index (BMI), and bioelectrical impedance analysis (BIA). Tools that are more direct, yet non-invasive include imaging techniques such as magnetic resonance imaging (MRI), DeXA, and computed tomography (CT) scan, as well as anthropometrics to assess leg and arm circumferences and skin fold thickness.

4.1 Biochemical Markers

Certain biological molecules, assessed in venous blood serum, have been mentioned as possible biochemical markers to detect decreased muscle mass. These include elevated expression of activin A (>0.35 ng/ml) (Ding et al. 2016) and myostatin (Wang and Mitch 2014; Hittel et al. 2009), and a decreased (<4.2 ng/mL) level of N-terminal peptide of procollagen type III (P3NP) (Fragala et al. 2014). Although, these molecules are being studied for their ability to indicate reduced muscle mass, current data suggest that it is premature to recommend their use in daily practice (Beaudart et al. 2016).

4.2 Imaging Assessment

Imaging is considered a valid tool, yet it is expensive and not always accessible to assess muscle mass. Techniques include MRI, DeXA, and CT. Imaging studies allow the assessment of body composition to be included in standard clinical or supportive care (Mitsiopoulos et al. 1998).

4.2.1 Dual-Energy X-Ray Absorptiometry (DeXA)

DeXA is a low-radiation technique to measure and estimate appendicular skeletal lean mass. Yet it cannot assess intra-muscular fat (Cruz-Jentoft et al. 2014; Levine et al. 2000). DeXA scan can only produce two-dimensional images and therefore cannot distinguish between subcutaneous and visceral adipose tissue.

Measuring appendicular skeletal lean mass can provide the ratio of individual body composition by calculating the sum of the non-fat and non-bone mass of the four limbs. As proposed by the European Society of Parenteral and Enteral Nutrition Special Interest Group (Cruz-Jentoft et al. 2014), two standard deviations below the mean level of young individuals (matched for age and gender) is considered a cut-off point for low muscle mass due to sarcopenia (e.g., appendicular mass relative to height squared, which is ≤ 7.23 kg/m^2 in men and ≤ 5.67 kg/m^2 in women). DeXA and BMI screening for muscle mass provide valid information. They were found to be very sensitive for estimating appendicular skeletal muscle mass of the lower limbs. Adjustment of anthropometric measurements for age, sex, or BMI could provide a better correlation with DeXA-measured lean mass (Dunsky et al. 2014).

4.2.2 Magnetic Resonance Imaging (MRI)

To determine whole body composition, MRI is used to provide valid, quantified volumetric information about soft and hard tissues such as muscle, subcutaneous adipose, intermuscular adipose and bone (Buford et al. 2012). The main advantages of MRI are that it provides excellent spatial resolution and differentiates body mass composition without radiation exposure. However, as with CT scan, DeXA, and BIA, MRI cannot analyze skeletal muscle quality. Yet it is more expensive compared to the other methods and its use is limited by local availability and technical expertise.

4.2.3 Computed Tomography (CT)

CT scan for assessing body composition has been shown to be superior to DeXA (Levine et al. 2000). CT can provide information on specific lean body mass, total body lean body mass, subcutaneous volume, visceral fat mass, total fat mass, subcutaneous fat-to-muscle ratio, and visceral-to-subcutaneous adipose tissue ratio (Mourtzakis et al. 2008). Major advantages of CT scans are high accuracy and reproducible results. Disadvantages, however, include radiation exposure and a higher cost compared to BIA and DeXA.

4.2.4 Bioelectrical Impedance Analysis (BIA)

BIA is a widely-available, inexpensive, user-friendly method that estimates the volume of fat and lean body mass based on the relation between the volume of a conductor and its electrical resistance. BIA involves placing electrodes on the hand and foot and measuring the impedance of a low-level electric current. The impedance is higher for fat and bone compared with soft tissue. Impedance can be affected by hydration status and fluid intake. Measurement accuracy and reference values have been established for older individuals (Reiss et al. 2016; Kim et al. 2015; Kim and Kim 2013).

5 Muscle Strength Assessment

The physiological meaning of reduced muscle strength is a loss of force-generating capacity. Usually, this loss is not correlated with morphological changes. In sarcopenia, mass reduction occurs to a greater extent and faster than strength reduction does. In clinical settings, hand or leg dynamometers are widely used for measuring muscle strength.

Isokinetic machines are used for research purposes, and in many cases isokinetic knee extensor torque is evaluated at 60°, 90°, and 120° (Bottaro et al. 2005). Isometric leg extension torque is well-correlated with handgrip strength and functional reach test (Jenkins et al. 2014). As with anthropometrics, this is easy to perform, inexpensive and does not require special training.

Isometric leg extension, to measure quadriceps femoris muscle strength, should be done while seated on a standard chair with the leg placed at 45° in knee extension. The manual muscle tester (MMT) is an ergonomic handheld device for objectively quantifying muscle strength. The test is performed with the clinician applying force to the patient's limb. The clinical objective of the test is to overcome or 'break' the patient's resistance. The MMT records the peak force and the time required to achieve the 'break' point, while providing reliable, accurate, and stable muscle strength readings that conform to most manual muscle testing protocols.

Hand grip strength should be measured while the patient is seated on a standard chair with the elbow flexed at 90°. Two measures should be taken with each arm, while the individual is encouraged to squeeze as hard as possible for 3–4 s for each trial. The higher of the two measurements is recorded. There are several types of hydraulic devices, such as the Jamar dynamometer (Patterson Medical, Warrenville, IL), which is the gold standard for this measurement, and pneumatic dynamometers such as the Martin Vigorimeter (Gebrüder Martin GmbH & Co., Tuttlingen, Germany) which is mostly used for patients with hand deformity (e.g., rheumatoid arthritis).

6 Sarcopenia – Functional Assessment

Functional performance is an integrative outcome of the overall effect of health and it reflects the ability of an individual to perform the physical tasks necessary for activities of daily living (Rosen and Reuben 2011). The Katz Index is widely used in community living and skilled nursing facilities to assess ability to dress, shower/bathe, sit down and rise from a chair, eat, and walk indoors (Katz 1983). The maximum score is 15, with a score of five indicating no functional limitations, a score of 6–10 indicating some functional limitations, and a score of 11 or more indicating several functional limitations.

Gait speed measurement is also frequently used in outpatient clinics. No special equipment is required, as it only needs a stop watch and a flat surface. References are suggested by the EWGSOP (Chiles Shaffer et al. 2016). In the 4-min gait speed test, men and women with a gait speed <0.8 m/s are described as having poor physical performance (Cuesta et al. 2015). The International Working Group on Sarcopenia has indicated that a diagnosis of sarcopenia is consistent with a gait speed of less than 1 m/s, or less than 400 m during a 6-min walking test (Fielding et al. 2011). There are several gait speed tests to use for sarcopenic individuals, but the Timed Up and Go (TUG) test (Bijlsma et al. 2014), the Short Physical Performance Battery (SPPB) (Steffl et al. 2016), and the five times sit-to-stand test (FTSST) (Lord et al. 2002) are the most widely used. These tests best correlate with mobility and disability. In the TUG test, individuals are asked to rise from a standard armchair, walk to a marker 3 m away, turn and walk back, and sit down again. The SPPB is a 10-min test with a maximum score of 12 points. It assesses gait speed (over 3–4 m) and individuals with a score ≤ 8 are characterized as having poor physical performance. The five times sit-to-stand test (FTSST) provides a reliable and valid indication of lower body strength and is commonly used. This timed test requires participants to rise from an armless chair, 43 cm high without using their arms and return to the seated position, five consecutive times. The test begins when the participant stands up from the initial sitting position at the go command and ends when the participant is in the final fully upright position at the end of the fifth stand.

A Japanese study has developed a screening tool to diagnose sarcopenia. The model is based on gender, demographic variables, blood profile especially albumin level, chronic diseases, physical activity information and anthropometrics (Ishii et al. 2014).

6.1 Sarcopenia Management

Patient-centered care has a key role in the management of sarcopenia. Since sarcopenia is frequently found in association with co-morbidities, e.g., osteoporosis, type II diabetes mellitus, chronic heart failure, poor balance, etc., treatment of these conditions is indispensable if the management of sarcopenia is to succeed. The intervention should involve a combination of physical exercises, dietary regimen, and nutritional supplements.

6.1.1 Physical Exercises (PE)

The PE and progressive resistance training have a strong effect on muscle strength, muscle mass, and physical performance in older people (Reid et al. 2015). There are no specific exercise protocols designed for individuals with sarcopenia. Therefore, general recommendations and guidelines for PE for elderly people suggested by WHO (2017) can serve as an initial protocol. General recommendations regarding the PE intervention in older people have also been suggested by the Asian Working Group for Sarcopenia (AWGS) (Chen et al. 2016b). To improve muscle function, the intervention should be for at least 3 months. Supervised resistance exercise or combined exercise programs should be recommended for sarcopenic or sedentary community-dwelling people. Progressive resistance, aerobic training predominantly effects muscle mass and muscle strength. Yet endurance aerobic training also is crucial to improve the function of the capillary bed in and around the muscle fibers and to increase local and systemic circulation. To improve muscle function, aerobic training should be performed least three times a week, for at least 150 min weekly, and for 12 consecutive weeks. Supervised, progressive resistance exercise should be performed 2–3 times a week, for a minimum of 30 min per week, for at least 3 months (Shad et al. 2016).

6.1.2 Diet and Nutrition

The European Union Geriatric Medicine Society (EUGMS), in cooperation with the dietary protein aging study group (PROT-AGE Study Group), has recently published nutritional recommendations for individuals with sarcopenia (Bauer et al. 2013). In general, some observational studies suggest that adequate protein intake (0.8–1.2 g/kg/day) and other dietary supplements (e.g., long-chains (omega−3 and 6) polyunsaturated fatty acids (PUFAs)) (Da Boit et al. 2017), β-hydroxy β-methylbutyrate (HMB), creatine, and vitamin D, combined with resistance exercise, may help preserve muscle mass in healthy older people. Supplementation with creatine, protein, or leucine, combined with exercise, seems to have a positive influence on physical performance (Martone et al. 2015).

A meta-analysis on diet has suggested that vitamin D supplementation could increase lower limb muscle strength (Stockton et al. 2011). A diet rich in protein or protein supplementation, mainly 60–90 min after physical exercise, accelerates muscle absorption of amino acids, especially leucine (Moore and Soeters 2015). Physical activity performed in the evening expands the overnight muscle protein synthetic response to pre-sleep protein absorption and permits more of amino acids to be used for *de novo* muscle protein synthesis during overnight sleep in older men (Holwerda et al. 2016).

7 Role of Clinicians and Primary Care Physicians

Despite that the definition and assessment of sarcopenia still lacks consensus, primary care physicians should consider a diagnosis of sarcopenia in older individuals (>65 years) with risk factors (e.g., diabetes, cancer, cardiovascular and pulmonary disease, osteoporosis, and poor balance). The Charlson Comorbidity Index (CCI) can be used (Perkins et al. 2004). The CCI uses a weighted scoring system (1–6 points) based on the presence of comorbid diseases. For instance, myocardial infarction or peripheral vascular disease equals 1 point, diabetes mellitus equals 2 points, liver disease 3 points, and cancer or acquired immune deficiency syndrome (AIDS) is 6 points.

When considering preventive health interventions, family physicians, geriatric physicians, and clinical dieticians should address comorbidities, functional status, activity level and risk factors. They should also check and inquire about caloric intake, protein quality, minerals (e.g., iron, magnesium), and serum vitamin B and vitamin D levels. The nutritional risk assessment can be carried out with the Mini Nutritional Assessment Short Form. The values range from 0 to 14 points and scores <11 identify patients at risk of poor nutrition.

8 Frailty Definition, Assessment, and Management

The definition of frailty is even more complex, with less consensus than sarcopenia. When Fried et al. (2001) have described the frailty phenotype and its association with mortality and morbidity, they noted a potential link between frailty and sarcopenia. However, frailty definitions vary, and it is largely conceptualized as increased vulnerability across multiple systems (Sirola et al. 2011; Rockwood and Mitnitski 2007). Based on an expert opinion statement (Rodríguez-Mañas et al. 2013), seven variables have been selected to define frailty: polypharmacy (Gnjidic et al. 2012), chronic heart failure (Phan et al. 2008), diabetes mellitus (Bourdel-Marchasson and Berrut 2005), subjective self-health assessment (Theou et al. 2015), physical activity questionnaire (Santos et al. 2015), mini-mental status evaluation (MMSE) (Bieniek et al. 2016), general health questionnaire (GHQ) (Kahlon et al. 2015), and a questionnaire to measure mood or depression (Bielderman et al. 2013). The variables outlined above are detailed below.

Polypharmacy, excluding nutraceuticals, is defined as the use of three or more prescribed medications. Chronic heart failure is reported as an independent predictor for frailty and is associated with other morbidities. Diabetes as a metabolic condition is associated with frailty. A subjective health assessment is widely used to assess frailty. This questionnaire addresses two main issues: (1) how is your health generally (very good/good/not so good/not good at all/bad) and (2) how is your health today as compared to your health a year ago (better/the same/not as good). Physical activity is one of the most powerful predictors for disability in daily living. The questionnaire on physical activity provides information about physical habits, frequency, duration, and average length of activity sessions. The MMSE is used to evaluate cognitive function ability. The maximum score is 30 and less than 24 points is considered a deficient cognition. The GHQ is mostly used to assess general health and also mood or depression. A score < 4 indicates no mood disturbance, 4–8 indicates mild disturbance, and 8–12 points a significant disturbance.

8.1 Frailty Assessment

Frailty is measurable. It is usually assessed late in life, in particular to evaluate the need for immediate care, help, or rehabilitation. With the worldwide increase in life expectancy, early detection of individuals at-risk may initiate an action to deter this health condition. To-date, comprehensive geriatric assessment (CGA) appears to be the most evidence-based process to detect and assess frailty (Chen et al. 2016a). Thus, CGA can provide a better understanding of the complex nature of frailty and support the development of healthcare practices to improve outcomes.

Frailty is mostly manifested by a low body weight, with a low hematocrit and serum albumin level < 3.4 g/dL (Blodgett et al. 2016). Several indices are commonly used to assess frailty. Woo et al. (2015) have suggested the FRAIL scale as the initial approach in detecting frailty in the community, enabling the targeted intervention to retard decline and future disability. The Fried Index (Fried et al. 2001) incorporates self-reported data of five criteria: unintended weight loss, exhaustion, leisure time activity, and some physical tests like hand grip strength. The Gill Frailty Index (Searle et al. 2008; Rothman et al. 2008) focuses entirely on a lower body physical performance measuring physical ability by sit-to-stand test and 20 m walking speed. This index, contrary to the Fried index, only consists of observed and measured physical performance tests. The main advantages of these instruments are that a single clinician can administer them with minimal safety concerns. The time to complete and score the Fried Index is 15–20 min and that for the Gill Index is less than 2 min. Kim et al. (2014) have reported an interesting observation on these two instruments. Individuals who were frail according to Gill Index may meet the Fried frailty criteria. Thus, a valid and reliable tool for frailty assessment is still an open issue that awaits a more comprehensive approach.

The Groningen Frailty Indicator (GFI) (Steverink et al. 2001) is a 15-item screening instrument that assesses frailty among home-dwelling elderly populations. It includes a combination of a professional and self-assessment questionnaire, addressing items that assess disability and can predict poor outcomes. It incorporates grades of frailty. The GFI is widely used in clinical practice and in clinical studies. A score of 4 or higher out of the 15 items represents moderate-to-severe frailty.

8.2 Frailty Management

Caring for frail elderly people is complicated and when they become ill it is even a more challenging and frustrating task. The elderly's illnesses are often nonspecific, unrecognized, and poorly documented. Treatment is complex and success is often unclear. Moreover, frailty status often requires end of life and sometimes palliative care. As noted above, frail people are vulnerable to common stressors, tend to have multiple interrelated medical and health problems, underweight, muscle weakness, impaired function, and high risk for falls and fractures. Therefore, the magnitude of their needs and the complexity of their health issues, require special focus and skills, along with a comprehensive, systematic, geriatric assessment to meet the challenges presented.

The basic concept of appropriate frailty management is that the clinician should move away from organs' impairment and pathology-based approaches toward the International Classification of Functioning, Disability and Health (ICF) approach, and thus include biopsychosocial aspects, as recommended by the French Society of Geriatrics and Gerontology (Rolland et al. 2011). The initial treatment goal is optimal management of any underlying illness or poor health condition that may increase frailty. Complex alterations of pharmacokinetics and pharmacodynamics occur in frail elderly patients (Cesari et al. 2015). Thus, recommended approach is of a kind 'to start low and go slow'. Another important challenge is monitoring of the progress of interventions, particularly in frail people with unusual disease appearances. Simple tracking of activity level in the frail elderly may however be helpful since when a frail person gets better he becomes more mobile and when the situation worsens less mobile.

8.3 Nutrition for Frailty

Although nutrition is considered a major factor in managing frailty, evidence of the effect of nutrition is often derived from short-term studies in selected samples; large clinical trials are lacking. Currently, there is no robust evidence for nutritional recommendations for individuals with frailty. To gain weight, a daily diet approach should be first to maintain balanced nutrient intake (Solon-Biet et al. 2015). Clinical studies show that a low-carbohydrate diet is beneficial for human health (Rosedale et al. 2009). A high-fat diet is associated with increased mortality and increased incidence of many metabolic diseases, including sarcopenic obesity, type II diabetes and cardiovascular problems (Baulderstone et al. 2012; Schrager et al. 2007). On the other hand, diets rich in unsaturated fatty acids lead to reduced blood levels of harmful low-density lipoproteins and increase the level of protective high-density lipoproteins (Mensink et al. 2003). Moreover, diets rich in natural, unsaturated fatty acids lower blood pressure, improve insulin sensitivity, and reduce the risks of cardiovascular and metabolic diseases (Da Silva et al. 2015). The Nutritional Geometric Framework (NGF) indicates the effects of a nature of various nutrient dimensions on lifespan and mortality (Raubenheimer et al. 2016). However, NGF has mostly been applied to the influence of protein intake relative to carbohydrates. Even if the results of randomized controlled trials are inconsistent regarding the effects of protein supplementation on physical function, several observational studies have suggested that maintaining adequate protein intake may help preserve energy in older people (Beck et al. 2016; Suominen et al. 2015). The frail elderly who have acute or chronic diseases need a higher dietary protein intake in a range of 1.2–1.5 g/kg/day (Chang 2017).

Hormone replacement therapy with testosterone, dehydroepiandrosterone (DHEA), or growth hormone has not proven beneficial (Morley and Malmstrom 2013; Lunenfeld 2006; Morley et al. 2005). Some drugs, which may be beneficial for treatment of fatigue, such as amantadine, methylphenidate, and modafinil, require further evidence-based evaluation and exploration (Mücke et al. 2016).

9 Relationship between Frailty and Sarcopenia

Sarcopenia and frailty often co-exist and both entail a physical function impairment as a core component. As indicated, numerous tools and instruments are available to assess sarcopenia. However, it is important for the clinician to be aware of which health aspects are being measured when assessing a person for frailty. This review stresses the importance of distinguishing and appropriately assessing older people who are frail from those who suffer from sarcopenia. The British Geriatrics Society believes that older people should be assessed for frailty and/or sarcopenia during all interactions with health and social care professionals (Turner et al. 2014). This would allow health care professionals to examine the benefits and risks of interventions and to allow individuals and their care givers to make rational decisions about the factors affecting their health.

Frailty, like sarcopenia, has been used interchangeably with disability (Rockwood et al. 2000). Yet despite the overlap in symptoms, frailty may be a universal condition of whole body wasting, including underweight and weakness, that directly affects functioning and recovery. Sarcopenia, on the other hand, is a more age-related, slowly progressive decline in muscle mass and strength, moderately affecting daily living activities. Frailty can be reversible, but this is not true for sarcopenia, which is a normal physiological change. Frailty can be understood as a continuum with two modifiable phases. Non-frail people can become pre-frail, and later on they become frail. There is a reverse way possible from frailty to pre-frailty status, and even potentially to a non-frail status. Frailty can be described as an 'acute' body condition, whereas sarcopenia is a normal, irreversible process resulting in programmed muscle death, i.e., muscle apoptosis. Therefore, even though, e.g., if an 80-year-old person in a pre-frailty condition, deteriorated to frailty, this could be halted or reversed, whereas for a person of the same age who already has some degree of sarcopenia, it is almost impossible to halt the physiological deterioration due to apoptosis.

Although a link between frailty and sarcopenia has been noted in a few studies (Mijnarends et al. 2015; Garatachea and Lucia 2013), this review focuses on the association of frailty and functional impairment and of sarcopenia and functional impairment. Yet both conditions result in disability in later life. Although low muscle mass increases the risk for fatigability, and low muscle strength and endurance increase functional disability, frailty increases the risk for underweight, functional disability, and other outcomes such as osteoporosis (Rosen and Klibanski 2009). It is well-documented that frail older adults are vulnerable, with minimal reserve capacity, and increased risk for malnutrition, institutionalization, and death (Payette et al. 2000). Sarcopenic older adults mostly demonstrate increased risk for gait, balance deficits, and falls. For these reasons, frailty care has mostly focused on nutrition and weight gain, whereas sarcopenia research has concentrated on physical interventions to increase muscle mass and strength. The management is different for frail *versus* sarcopenic older adults. Frail individuals should consume calories to increase body weight, sarcopenic individuals should increase their physical activity by performing resistance and endurance exercises. To gain weight, frail people's diet should include carbohydrate, lipids, and proteins. In addition, several studies on dietary patterns have shown positive effects of enhanced nutritional intake also on general health, e.g., lower blood pressure, reduced risk of coronary heart failure, reduced risk of type 2 diabetes, and cancers (Asp and Bryngelsson 2008). Sarcopenic people who need to gain muscle mass should consume protein, particularly 1.5–2 h after performing physical exercises.

Frail people often demonstrate low body weight or low BMI (<25 kg/m^2), despite evidence of abdominal obesity (Buch et al. 2016), with a greater likelihood of functional limitations and difficulties performing daily living activities (Byard 2015). The mechanism of interaction of the frail state with obesity has not been clearly defined, but it might involve hormonal dysregulation and inflammatory pathways, as well as oxidative stress. Insulin resistance, associated with abdominal obesity, may promote abnormal 'colonization' and emergence of ectopic fat in muscle, which is associated with functional limitations (Auyeung et al. 2013). The Foundation for the National Institutes of Health Sarcopenia Project validated cut points and reported that older adults with sarcopenia have a low-to-moderate BMI, which appears to protect against functional limitations (Batsis et al. 2015).

10 Clinical and Practical Recommendations

- I propose the assessment of muscle mass primarily with dual-energy X-ray absorptiometry (DeXA). In case the technique is unavailable, anthropometry can be easily used in primary care settings as an initial screening tool for patients with low muscle mass. Patients should then be referred for further evaluation in clinical specialty settings.
- Physical performance should be primarily assessed by measuring gait speed. The Short Physical Performance Battery (SPPB) test might be limited by administration time, but might also be useful to identify men and women with low physical performance.
- A comprehensive geriatric assessment (CGA) appears the most evidence-based process for detecting and assessing frailty.
- The Groningen Frailty Indicator, the Fried Index, and the Gill Frailty Index are valid and reliable tools for assessing and monitoring frailty.
- Whereas further studies are required to provide a full evidence-based guidance to clinicians, current management should include physical activity advice, particularly progressive resistance training, treatment and prevention of vitamin D deficiency, and adequate and balanced energy intake emphasizing dietary protein.
- Emphasis on the importance of education and increased awareness of clinicians concerning the potential deleterious effects of frailty and sarcopenia.
- Careful attention given to individuals older than 65 years with low BMI or abdominal obesity.

11 Summary and Conclusions

This review describes differences, similarities, and commonalities in the definition, assessment, and management of frailty and sarcopenia. Despite the similarities between the two, frailty and sarcopenia are two separate health conditions. Therefore, it is important to diagnose and manage them as separate entities.

- The definition of sarcopenia is agreed upon by the European Working Group on Sarcopenia in Older People (gait speed, handgrip strength, and muscle mass), and the International Working Group on Sarcopenia (gait speed and muscle mass).
- The consensus definition of frailty include the Fried phenotype (weight loss, exhaustion, physical inactivity, handgrip strength, and walk time) (Fried et al. 2001) and the Rockwood phenotype (use of walking aid, activities of daily living, incontinence, and cognitive impairment) (Rockwood et al. 2000).
- Primary sarcopenia is age-related, but it is frequently found in association with comorbidities, such as osteoporosis, malnutrition, and type 2 diabetes mellitus. Thus, it should be considered a consequence of the coexisting pathological conditions, i.e., 'secondary sarcopenia'.
- Several tools are currently available for measuring muscle mass, muscle strength, and physical performance, which are of use for the diagnosis and follow-up of sarcopenia. However, these tools remain to be fully adopted for widespread use in clinical daily practice.
- The development of pharmaceutical therapies for sarcopenia and frailty has been delayed, in part because of the lack of consensus regarding the definitions of the two conditions.

Physicians and other healthcare professionals have an important role to play in the assessment and management of sarcopenia to reduce its impact on individuals' well-being, the development of disability, and on health resource utilization. This review suggests that frailty and sarcopenia differentially affect functional capability, morbidity, quality of life, and disability in daily living activities. Frailty is characterized by low weight and reflects nutritional status, whereas sarcopenia – the loss of muscle mass – is more accurate and can be a quantitative, global marker of frailty. This review also highlights the importance of frailty and sarcopenia in predicting post-operative outcomes among individuals undergoing surgery for cancer.

Conflicts of Interest The author declares no conflicts of interest in relation to this article.

References

Araujo AB, Chiu GR, Kupelian V, Hall SA, Williams RE, Clark RV, McKinlay JB (2010) Lean mass, muscle strength, and physical function in a diverse population of men: a population-based cross-sectional study. BMC Public Health 10:508

Asp NG, Bryngelsson S (2008) Health claims in Europe: new legislation and PASSCLAIM for substantiation. J Nutr 138(6):1210S–1215S

Auyeung TW, Lee JS, Leung J, Kwok T, Woo J (2013) Adiposity to muscle ratio predicts incident physical limitation in a cohort of 3,153 older adults – an alternative measurement of sarcopenia and sarcopenic obesity. Age (Dordr) 35(4):1377–1385

Azzopardi RV, Vermeiren S, Gorus E, Habbig AK, Petrovic M, Van Den Noortgate N, De Vriendt P, Bautmans I, Beyer I, Gerontopole Brussels Study Group (2016) Linking frailty instruments to the international classification of functioning, disability, and health: a systematic review. J Am Med Dir Assoc 17 (11):1066.e1–1066.e11

Barbat-Artigas S, Rolland Y, Zamboni M, Aubertin-Leheudre M (2012) How to assess functional status: a new muscle quality index. J Nutr Health Aging 16 (1):67–77

Batsis JA, Mackenzie TA, Lopez-Jimenez F, Bartels SJ (2015) Sarcopenia, sarcopenic obesity, and functional impairments in older adults: national health and nutrition examination surveys 1999–2004. Clin Nutr 35 (12):1031–1039

Bauer J, Biolo G, Cederholm T, Cesari M, Cruz-Jentoft AJ, Morley JE, Phillips S, Sieber C, Stehle P, Teta D, Visvanathan R, Volpi E, Boirie Y (2013) Evidence-based recommendations for optimal dietary protein intake in older people: a position paper from the PROT-AGE study group. J Am Med Dir Assoc 14 (8):542–559

Baulderstone L, Yaxley A, Luszcz M, Miller M (2012) Diet liberalization in older Australians decreases frailty without increasing the risk of developing chronic disease. J Frailty Aging 4:174–182

Beaudart C, McCloskey E, Bruyère O, Cesari M, Rolland Y, Rizzoli R, Araujo de Carvalho I, Amuthavalli Thiyagarajan J, Bautmans I, Bertière MC, Brandi ML, Al-Daghri NM, Burlet N, Cavalier E, Cerreta F, Cherubini A, Fielding R, Gielen E, Landi F, Petermans J, Reginster JY, Visser M, Kanis J, Cooper C (2016) Sarcopenia in daily practice: assessment and management. BMC Geriatr 16(1):170

Beck AM, Dent E, Baldwin C (2016) Nutritional intervention as part of functional rehabilitation in older people with reduced functional ability: a systematic review and meta-analysis of randomised controlled studies. J Hum Nutr Diet 29(6):733–745

Bielderman A, van der Schans CP, van Lieshout MR, de Greef MH, Boersma F, Krijnen WP, Steverink N (2013) Multidimensional structure of the Groningen frailty indicator in community-dwelling older people. BMC Geriatr 13:86

Bieniek J, Wilczyński K, Szewieczek J (2016) Fried frailty phenotype assessment components as applied to geriatric inpatients. Clin Interv Aging 11:453–459

Bijlsma AY, Meskers CG, van den Eshof N, Westendorp RG, Sipilä S, Stenroth L, Sillanpää E, McPhee JS, Jones DA, Narici MV, Gapeyeva H, Pääsuke M, Voit T, Barnouin Y, Hogrel JY, Butler Browne G, Maier AB (2014) Diagnostic criteria for sarcopenia and physical performance. Age (Dordr) 36 (1):275–285

Blodgett JM, Theou O, Howlett SE, Wu FC, Rockwood K (2016) A frailty index based on laboratory deficits in community-dwelling men predicted their risk of adverse health outcomes. Age Ageing 45(4):463–468

Bottaro M, Russo AF, de Oliveira RJ (2005) The effects of rest interval on quadriceps torque during an isokinetic testing protocol in elderly. J Sports Sci Med 4(3):285–290

Bourdel-Marchasson I, Berrut G (2005) Caring the elderly diabetic patient with respect to concepts of successful aging and frailty. Diabetes Metab 2:5S13–5S19

Bowen ME (2012) The relationship between body weight, frailty, and the disablement process. J Gerontol B Psychol Sci Soc Sci 67(5):618–626

Buch A, Carmeli E, Boker LK, Marcus Y, Shefer G, Kis O, Berner Y, Stern N (2016) Muscle function and fat content in relation to sarcopenia, obesity and frailty of old age – an overview. Exp Gerontol 76:25–32

Buford TW, Lott DJ, Marzetti E, Wohlgemuth SE, Vandenborne K, Pahor M, Leeuwenburgh C, Manini TM (2012) Age-related differences in lower extremity tissue compartments and associations with physical function in older adults. Exp Gerontol 47:38–44

Byard RW (2015) Forensic issues at the extremes of weight: from frailty syndrome through frail obesity to morbid obesity. J Forensic Legal Med 35:38–39

Cao L, Morley JE (2016) Sarcopenia is recognized as an independent condition by an international classification of disease, tenth revision, clinical modification (ICD-10-CM) code. J Am Med Dir Assoc 17 (8):675–677

Cesari M, Fielding R, Bénichou O, Bernabei R, Bhasin S, Guralnik JM, Jette A, Landi F, Pahor M, Rodriguez-Manas L, Rolland Y, Roubenoff R, Sinclair AJ, Studenski S, Travison T, Vellas B (2015) Pharmacological interventions in frailty and sarcopenia: report by the international conference on frailty and sarcopenia research task force. J Frailty Aging 4 (3):114–120

Chang SF (2017) Frailty is a major related factor for at risk of malnutrition in community-dwelling older adults. J Nurs Scholarsh 49(1):63–72

Chang SF, Yang RS, Lin TC, Chiu SC, Chen ML, Lee HC (2014) The discrimination of using the short physical performance battery to screen frailty for community-dwelling elderly people. J Nurs Scholarsh 46 (3):207–215

Chen LK, Hwang AC, Liu LK, Lee WJ, Peng LN (2016a) Frailty is a geriatric syndrome characterized by multiple impairments: a comprehensive approach is needed. J Frailty Aging 5(4):208–213

Chen LK, Lee WJ, Peng LN, Liu LK, Arai H, Akishita M, Asian Working Group for Sarcopenia (2016b) Recent advances in sarcopenia research in Asia: 2016 update from the Asian working Group for Sarcopenia. J Am Med Dir Assoc 17(8):767.e1–767.e7

Chiles Shaffer N, Ferrucci L, Shardell M, Simonsick EM, Studenski S (2016) Agreement and predictive validity using less-conservative foundation for the National Institutes of Health sarcopenia project weakness Cutpoints. J Am Geriatr Soc. doi:10.1111/jgs.14706

Chow WB, Rosenthal RA, Merkow RP, Ko CY, Esnaola NF (2012) Optimal preoperative assessment of the geriatric surgical patient: a best practices guideline from the American College of Surgeons National Surgical Quality Improvement Program and the American Geriatrics Society. J Am Coll Surg 215(4):453–466

Cruz-Jentoft AJ, Landi F, Schneider SM, Zúñiga C, Arai H, Boirie Y, Chen LK, Fielding RA, Martin FC, Michel JP, Sieber C, Stout JR, Studenski SA, Vellas B, Woo J, Zamboni M, Cederholm T (2014) Prevalence of and interventions for sarcopenia in ageing adults: a systematic review. Report of the international sarcopenia initiative (EWGSOP and IWGS). Age Ageing 43(6):748–759

Cuesta F, Formiga F, Lopez-Soto A, Masanes F, Ruiz D, Artaza I, Salvà A, Serra-Rexach JA, Rojano I, Luque X, Cruz-Jentoft AJ (2015) Prevalence of sarcopenia in patients attending outpatient geriatric clinics: the ELLI study. Age Ageing 44(5):807–809

Da Boit M, Sibson R, Sivasubramaniam S, Meakin JR, Greig CA, Aspden RM, Thies F, Jeromson S, Hamilton DL, Speakman JR, Hambly C, Mangoni AA, Preston T, Gray SR (2017) Sex differences in the effect of fish oil supplementation on the adaptive response to resistance exercise training in older people: a randomized control trial. Am J Clin Nutr 105 (1):151–158

Da Silva MS, Julien P, Pérusse L, Vohl MC, Rudkowska I (2015) Natural rumen-derived trans fatty acids are associated with metabolic markers of cardiac health. Lipids 50(9):873–882

Ding H, Zhang G, Sin KW, Liu Z, Lin RK, Li M, Li YP (2016) Activin a induces skeletal muscle catabolism via p38β mitogen-activated protein kinase. J Cachexia Sarcopenia Muscle. doi:10.1002/jcsm.12145

Drey M (2011) Sarcopenia – pathophysiology and clinical relevance. Wien Med Wochenschr 161 (17–18):402–408

Dunsky A, Zach S, Zeev A, Goldbourt U, Shimony T, Goldsmith R, Netz Y (2014) Level of physical activity and anthropometric characteristics in old age – results from a national health survey. Eur Rev Aging Phys Act 11(2):149–157

Fielding RA, Vellas B, Evans WJ, Bhasin S, Morley JE, Newman AB, Abellan van Kan G, Andrieu S, Bauer J, Breuille D, Cederholm T, Chandler J, De Meynard C, Donini L, Harris T, Kannt A, Keime Guibert F, Onder G, Papanicolaou D, Rolland Y, Rooks D, Sieber C, Souhami E, Verlaan S, Zamboni M (2011) Sarcopenia: an undiagnosed condition in older adults. Current consensus definition: prevalence, etiology, and consequences. International working group on sarcopenia. J Am Med Dir Assoc 12(4):249–256

Figueiredo CS, Assis MG, Silva SL, Dias RC, Mancini MC (2013) Functional and cognitive changes in community-dwelling elderly: longitudinal study. Brazil J Phys Ther 7(3):297–306

Fragala MS, Jajtner AR, Beyer KS, Townsend JR, Emerson NS, Scanlon TC, Oliveira LP, Hoffman JR, Stout JR (2014) Biomarkers of muscle quality: N-terminal propeptide of type III procollagen and C-terminal agrin fragment responses to resistance exercise training in older adults. J Cachex Sarcopenia Muscle 5(2):139–148

Fried LP, Tangen CM, Walston J, Newman AB, Hirsch C, Gottdiener J, Seeman T, Tracy R, Kop WJ, Burke G, MA MB, Cardiovascular Health Study Collaborative Research Group (2001) Frailty in older adults: evidence for a phenotype. J Gerontol A Biol Sci Med Sci 56:M146–M156

Garatachea N, Lucia A (2013) Genes, physical fitness and ageing. Ageing Res Rev 12(1):90–102

Gnjidic D, Hilmer SN, Blyth FM, Naganathan V, Waite L, Seibel MJ, McLachlan AJ, Cumming RG, Handelsman DJ, Le Couteur DG (2012) Polypharmacy cutoff and outcomes: five or more medicines were used to identify community-dwelling older men at risk of different adverse outcomes. J Clin Epidemiol 65(9):989–995

Hamerman D (1999) Toward an understanding of frailty. Ann Intern Med 130(11):945–950

Hassani A, Kubicki A, Brost V, Mourey F, Yang F (2015) Kinematic analysis of motor strategies in frail aged adults during the timed up and go: how to spot the motor frailty? Clin Interv Aging 10:505–513

Heuberger RA (2011) The frailty syndrome: a comprehensive review. J Nutr Gerontol Geriatr 30 (4):315–368

Hittel DS, Berggren JR, Shearer J, Boyle K, Houmard JA (2009) Increased secretion and expression of myostatin in skeletal muscle from extremely obese women. Diabetes 58(1):30–38

Holwerda AM, Kouw IW, Trommelen J, Halson SL, Wodzig WK, Verdijk LB, van Loon LJ (2016) Physical activity performed in the evening increases the overnight muscle protein synthetic response to pre-sleep protein ingestion in older men. J Nutr 146 (7):1307–1314

Inouye SK, Peduzzi PN, Robison JT, Hughes JS, Horwitz RI, Concato J (1998) Importance of functional measures in predicting mortality among older hospitalized patients. JAMA 279(15):1187–1193

Ishii S, Tanaka T, Shibasaki K, Ouchi Y, Kikutani T, Higashiguchi T, Obuchi SP, Ishikawa-Takata K, Hirano H, Kawai H, Tsuji T, Iijima K (2014) Development of a simple screening test for sarcopenia in older adults. Geriatr Gerontol Int 14(Suppl 1):93–101

Jenkins ND, Buckner SL, Bergstrom HC, Cochrane KC, Goldsmith JA, Housh TJ, Johnson GO, Schmidt RJ, Cramer JT (2014) Reliability and relationships among handgrip strength, leg extensor strength and power, and balance in older men. Exp Gerontol 58:47–50

Kahlon S, Pederson J, Majumdar SR, Belga S, Lau D, Fradette M, Boyko D, Bakal JA, Johnston C, Padwal RS, McAlister FA (2015) Association between frailty and 30-day outcomes after discharge from hospital. Can Med Assoc J 187(11):799–804

Katz S (1983) Assessing self-maintenance: activities of daily living, mobility, and instrumental activities of daily living. J Am Geriatr Soc 31(12):721–727

Kim M, Kim H (2013) Accuracy of segmental multi-frequency bioelectrical impedance analysis for assessing whole-body and appendicular fat mass and lean soft tissue mass in frail women aged 75 years and older. Eur J Clin Nutr 67:395–400

Kim H, Higgins PA, Canaday DH, Burant CJ, Hornick TR (2014) Frailty assessment in the geriatric outpatient clinic. Geriatr Gerontol Int 14(1):78–83

Kim M, Shinkai S, Murayama H, Mori S (2015) Comparison of segmental multifrequency bioelectrical impedance analysis with dual-energy X-ray absorptiometry for the assessment of body composition in a community-dwelling older population. Geriatr Gerontol Int 15:1013–1022

Kulkarni B, Kuper H, Taylor A, Wells JC, Radhakrishna KV, Kinra S, Ben-Shlomo Y, Smith GD, Ebrahim S, Byrne NM, Hills AP (2013) Development and validation of anthropometric prediction equations for estimation of lean body mass and appendicular lean soft tissue in Indian men and women. J Appl Physiol 115:1156–1162

Landi F, Russo A, Liperoti R, Pahor M, Tosato M, Capoluongo E, Bernabei R, Onder G (2010) Midarm muscle circumference, physical performance and mortality: results from the aging and longevity study in the Sirente geographic area (ilSIRENTE study). Clin Nutr 29:441–447

Landi F, Onder G, Russo A, Liperoti R, Tosato M, Martone AM, Capoluongo E, Bernabei R (2014) Calf circumference, frailty and physical performance among older adults living in the community. Clin Nutr 33(3):539–544

Lang PO, Michel JP, Zekry D (2009) Frailty syndrome: a transitional state in a dynamic process. Gerontology 55(5):539–549

Levine JA, Abboud L, Barry M, Reed JE, Sheedy PF, Jensen MD (2000) Measuring leg muscle and fat mass in humans: comparison of CT and dual-energy X-ray absorptiometry. J Appl Physiol 88:452–456

Lord SR, Murray SM, Chapman K, Munro B, Tiedemann A (2002) Sit-to-stand performance depends on sensation, speed, balance, and psychological status in addition to strength in older people. J Gerontol A Biol Sci Med Sci 57(8):M539–M543

Lunenfeld B (2006) Endocrinology of the aging male. Minerva Ginecol 58(2):153–170

Malatesta M, Cardani R, Pellicciari C, Meola G (2014) RNA transcription and maturation in skeletal muscle cells are similarly impaired in myotonic dystrophy and sarcopenia: the ultrastructural evidence. Front Aging Neurosci 6:196

Martone AM, Lattanzio F, Abbatecola AM, Carpia DL, Tosato M, Marzetti E, Calvani R, Onder G, Landi F (2015) Treating sarcopenia in older and oldest old. Curr Pharm Des 21(13):1715–1722

McCleary NJ, Dotan E, Browner I (2014) Refining the chemotherapy approach for older patients with colon cancer. J Clin Oncol 32:2570–2580

Mensink RP, Zock PL, Kester AD, Katan MB (2003) Effects of dietary fatty acids and carbohydrates on the ratio of serum total to HDL cholesterol and on serum lipids and apolipoproteins: a meta analysis of 60 controlled trials. Am J Clin Nutr 77(5):1146–1155

Mijnarends DM, Schols JM, Meijers JM, Tan FE, Verlaan S, Luiking YC, Morley JE, Halfens RJ (2015) Instruments to assess sarcopenia and physical frailty in older people living in a community (care) setting: similarities and discrepancies. J Am Med Dir Assoc 16(4):301–308

Mitsiopoulos N, Baumgartner RN, Heymsfield SB, Lyons W, Gallagher D, Ross R (1998) Cadaver validation of skeletal muscle measurement by magnetic resonance imaging and computerized tomography. J Appl Physiol 85:115–122

Moore DR, Soeters PB (2015) The biological value of protein. Nestle Nutr Inst Workshop Ser 82:39–51. doi:10.1159/000382000

Morley JE, Malmstrom TK (2013) Frailty, sarcopenia, and hormones. Endocrinol Metab Clin N Am 42 (2):391–405

Morley JE, Kim MJ, Haren MT (2005) Frailty and hormones. Rev Endocr Metab Disord 6(2):101–108

Mourtzakis M, Prado CM, Lieffers JR, Reiman T, McCargar LJ, Baracos VE (2008) A practical and precise approach to quantification of body composition in cancer patients using computed tomography

images acquired during routine care. Appl Physiol Nutr Metab 33(5):997–1006
Mücke M, Mochamat CH, Peuckmann-Post V, Minton O, Stone P, Radbruch L (2016) Pharmacological treatments for fatigue associated with palliative care: executive summary of a Cochrane collaboration systematic review. J Cachex Sarcopenia Muscle 7 (1):23–27
Payette H, Coulombe C, Boutier V, Gray-Donald K (2000) Nutrition risk factors for institutionalization in a free-living functionally dependent elderly population. J Clin Epidemiol 53(6):579–587
Perkins AJ, Kroenke K, Unützer J, Katon W, Williams JW Jr, Hope C, Callahan CM (2004) Common comorbidity scales were similar in their ability to predict health care costs and mortality. J Clin Epidemiol 57 (10):1040–1048
Peters LL, Boter H, Buskens E, Slaets JP (2012) Measurement properties of the Groningen frailty indicator in home-dwelling and institutionalized elderly people. J Am Med Dir Assoc 13(6):546–5451
Phan HM, Alpert JS, Fain M (2008) Frailty, inflammation, and cardiovascular disease: evidence of a connection. Am J Geriatr Cardiol 17(2):101–107
Raîche M, Hébert R, Dubois MF (2008) PRISMA-7: a case-finding tool to identify older adults with moderate to severe disabilities. Arch Gerontol Geriatr 47 (1):9–18
Raubenheimer D, Simpson SJ, Le Couteur DG, Solon-Biet SM, Coogan SC (2016) Nutritional ecology and the evolution of aging. Exp Gerontol 86:50–61
Ravaglia G, Forti P, Lucicesare A, Pisacane N, Rietti E, Bianchin M, Dalmonte E (2008) Physical activity and dementia risk in the elderly: findings from a prospective Italian study. Neurology 70(19 Pt 2):1786–1794
Reid KF, Martin KI, Doros G, Clark DJ, Hau C, Patten C, Phillips EM, Frontera WR, Fielding RA (2015) Comparative effects of light or heavy resistance power training for improving lower extremity power and physical performance in mobility-limited older adults. J Gerontol A Biol Sci Med Sci 70(3):374–380
Reiss J, Iglseder B, Kreutzer M, Weilbuchner I, Treschnitzer W, Kässmann H, Pirich C, Reiter R (2016) Case finding for sarcopenia in geriatric inpatients: performance of bioimpedance analysis in comparison to dual X-ray absorptiometry. BMC Geriatr 16:52
Rockwood K, Mitnitski A (2007) Frailty in relation to the accumulation of deficits. J Gerontol A Biol Sci Med Sci 62(7):722–727
Rockwood K, Hogan DB, MacKnight C (2000) Conceptualisation and measurement of frailty in elderly people. Drugs Aging 17(4):295–302
Rodríguez-Mañas L, Féart C, Mann G, Viña J, Chatterji S, Chodzko-Zajko W, Gonzalez-Colaço Harmand M, Bergman H, Carcaillon L, Nicholson C, Scuteri A, Sinclair A, Pelaez M, Van der Cammen T, Beland F, Bickenbach J, Delamarche P, Ferrucci L, Fried LP, Gutiérrez-Robledo LM, Rockwood K, Rodríguez Artalejo F, Serviddio G, Vega E, FOD-CC Group (Appendix 1) (2013) Searching for an operational definition of frailty: a Delphi method based consensus statement: the frailty operative definition-consensus conference project. J Gerontol A Biol Sci Med Sci 68(1):62–67
Rolland Y, Lauwers-Cances V, Cournot M, Nourhashémi F, Reynish W, Rivière D, Vellas B, Grandjean H (2003) Sarcopenia, calf circumference, and physical function of elderly women: a cross-sectional study. J Am Geriatr Soc 51:1120–1124
Rolland Y, Benetos A, Gentric A, Ankri J, Blanchard F, Bonnefoy M, de Decker L, Ferry M, Gonthier R, Hanon O, Jeandel C, Nourhashemi F, Perret-Guillaume C, Retornaz F, Bouvier H, Ruault G, Berrut G (2011) Frailty in older population: a brief position paper from the French society of geriatrics and gerontology. Geriatr Psychol Neuropsychiatr Vieil 9 (4):387–390
Rosedale R, Westman EC, Konhilas JP (2009) Clinical experience of a diet designed to reduce aging. J Appl Res 9(4):159–165
Rosen CJ, Klibanski A (2009) Bone, fat, and body composition: evolving concepts in the pathogenesis of osteoporosis. Am J Med 122(5):409–414
Rosen SL, Reuben DB (2011) Geriatric assessment tools. Mt Sinai J Med 78(4):489–497
Rosenberg IR (1989) Summary comments. Am J Clin Nutr 50:1231–1233
Rothman MD, Leo-Summers L, Gill TM (2008) Prognostic significance of potential frailty criteria. J Am Geriatr Soc 56(12):2211–2116
Saliba D, Elliott M, Rubenstein LZ, Solomon DH, Young RT, Kamberg CJ, Roth C, MacLean CH, Shekelle PG, Sloss EM, Wenger NS (2001) The vulnerable elders survey: a tool for identifying vulnerable older people in the community. J Am Geriatr Soc 49 (12):1691–1699
Santos PH, Fernandes MH, Casotti CA, Coqueiro Rda S, Carneiro JA (2015) The profile of fragility and associated factors among the elderly registered in a family health unit. Cien Saude Colet 20(6):1917–1924
Schrager MA, Metter EJ, Simonsick E, Ble A, Bandinelli S, Lauretani F, Ferrucci L (2007) Sarcopenic obesity and inflammation in the InCHIANTI study. J Appl Physiol (1985) 102 (3):919–925
Scott D, Trbojevic T, Skinner E, Clark RA, Levinger P, Haines TP, Sanders KM, Ebeling PR (2015) Associations of calf inter- and intra-muscular adipose tissue with cardiometabolic health and physical function in community-dwelling older adults. J Musculoskelet Neuronal Interact 15(4):350–357
Searle SD, Mitnitski A, Gahbauer EA, Gill TM, Rockwood K (2008) A standard procedure for creating a frailty index. BMC Geriatr 8:24
Shad BJ, Wallis G, van Loon LJ, Thompson JL (2016) Exercise prescription for the older population: the interactions between physical activity, sedentary

time, and adequate nutrition in maintaining musculoskeletal health. Maturitas 93:78–82

Silva AM, Shen W, Heo M, Gallagher D, Wang Z, Sardinha LB, Heymsfield SB (2010) Ethnicity-related skeletal muscle differences across the lifespan. Am J Hum Biol 22(1):76–82

Sirola J, Pitkala KH, Tilvis RS, Miettinen TA, Strandberg TE (2011) Definition of frailty in older men according to questionnaire data (RAND-36/SF-36): the Helsinki businessmen study. J Nutr Health Aging 15 (9):783–787

Solon-Biet SM, Mitchell SJ, de Cabo R, Raubenheimer D, Le Couteur DG, Simpson SJ (2015) Macronutrients and caloric intake in health and longevity. J Endocrinol 226(1):R17–R28

Steffl M, Musalek M, Kramperova V, Petr M, Kohlikova E, Holmerova I, Volicer L (2016) Assessment of diagnostics tools for sarcopenia severity using the item response theory (IRT). J Nutr Health Aging 20(10):1051–1055

Sternberg SA, Wershof Schwartz A, Karunananthan S, Bergman H, Mark Clarfield A (2011) The identification of frailty: a systematic literature review. J Am Geriatr Soc 59(11):2129–2138

Steverink N, Slaets J, Schuurmans H, Van Lis M (2001) Measuring frailty: developing and testing the GFI (Groningen frailty indicator). Gerontologist 41:236e237

Stockton KA, Mengersen K, Paratz JD, Kandiah D, Bennell KL (2011) Effect of vitamin D supplementation on muscle strength: a systematic review and meta-analysis. Osteoporos Int 22(3):859–871

Suominen MH, Puranen TM, Jyväkorpi SK, Eloniemi-Sulkava U, Kautiainen H, Siljamäki-Ojansuu U, Pitkalä KH (2015) Nutritional guidance improves nutrient intake and quality of life, and may prevent falls in aged persons with Alzheimer disease living with a spouse (NuAD trial). J Nutr Health Aging 19 (9):901–907

Tamura F, Kikutani T, Tohara T, Yoshida M, Yaegaki K (2012) Tongue thickness relates to nutritional status in the elderly. Dysphagia 27(4):556–561

Theou O, O'Connell MD, King-Kallimanis BL, O'Halloran AM, Rockwood K, Kenny RA (2015) Measuring frailty using self-report and test-based health measures. Age Ageing 44(3):471–477

Turner G, Clegg A, British Geriatrics Society, Age UK, Royal College of General Practioners (2014) Best practice guidelines for the management of frailty: a British geriatrics society, Age UK and Royal College of general practitioners report. Age Ageing 43 (6):744–747

Vermeulen J, Neyens JC, van Rossum E, Spreeuwenberg MD, de Witte LP (2011) Predicting ADL disability in community-dwelling elderly people using physical frailty indicators: a systematic review. BMC Geriatr 11:33. doi:10.1186/1471-2318-11-33

Vettor R, Milan G, Franzin C, Sanna M, De Coppi P, Rizzuto R, Federspil G (2009) The origin of intermuscular adipose tissue and its pathophysiological implications. Am J Physiol Endocrinol Metab 297 (5):E987–E998

Villani AM, Miller M, Cameron ID, Kurrle S, Whitehead C, Crotty M (2013) Body composition in older community-dwelling adults with hip fracture: portable field methods validated by dual-energy X-ray absorptiometry. Brit J Nutr 109(7):1219–1229

Villani AM, Crotty M, Cameron ID, Kurrle SE, Skuza PP, Cleland LG, Cobiac L, Miller MD (2014) Appendicular skeletal muscle in hospitalized hip-fracture patients: development and cross-validation of anthropometric prediction equations against dual-energy X-ray absorptiometry. Age Ageing 43:857–862

Visser M, Kritchevsky SB, Goodpaster BH, Newman AB, Nevitt M, Stamm E, Harris TB (2002) Leg muscle mass and composition in relation to lower extremity performance in men and women aged 70 to 79: the health, aging and body composition study. J Am Geriatr Soc 50:897–904

Walston J, Hadley EC, Ferrucci L, Guralnik JM, Newman AB, Studenski SA, Ershler WB, Harris T, Fried LP (2006) Research agenda for frailty in older adults: toward a better understanding of physiology and etiology: summary from the American Geriatrics Society/ National Institute on Aging research conference on frailty in older adults. J Am Geriatr Soc 54 (6):991–1001

Wang XH, Mitch WE (2014) Mechanisms of muscle wasting in chronic kidney disease. Nat Rev Nephrol 10(9):504–516

Wei N, Pang MY, Ng SS, Ng GY (2016) Optimal frequency/time combination of whole body vibration training for improving muscle size and strength of people with age-related muscle loss (sarcopenia): a randomized controlled trial. Geriatr Gerontol Int. doi:10.1111/ggi.12878

WHO (2015) Facts about ageing http://www.who.int/ageing/about/facts/en/. Accessed on 23 Jan 2017

WHO (2017) http://www.who.int/dietphysicalactivity/factsheet_olderadults/en/. Accessed on 23 Jan 2017

Woo J, Yu R, Wong M, Yeung F, Wong M, Lum C (2015) Frailty screening in the community using the FRAIL scale. J Am Med Dir Assoc 16(5):412–419

Xue QL (2011) The frailty syndrome: definition and natural history. Clin Geriatr Med 27(1):1–15

Yu S, Appleton S, Chapman I, Adams R, Wittert G, Visvanathan T, Visvanathan R (2015) An anthropometric prediction equation for appendicular skeletal muscle mass in combination with a measure of muscle function to screen for sarcopenia in primary and aged care. J Am Med Dir Assoc 16:25–30

Advs Exp. Medicine, Biology - Neuroscience and Respiration (2017) 32: 69–79
DOI 10.1007/5584_2017_3

Published online: 6 April 2017

Exercise Strategies to Counteract Brain Aging Effects

Dominika Szalewska, Marek Radkowski, Urszula Demkow, and Pawel J. Winklewski

Abstract

Stimulating structural and functional adaptation that improves cognitive performance in specific tasks is the major objective of therapeutic exercise training. In this review we briefly summarize central physiological mechanisms activated by exercise. We further discuss the influence of different kinds of exercise on cognitive improvement. In particular, the effects on cognitive function of aerobic endurance, resistance and respiratory exercise, and combinations thereof are presented. The accumulating evidence reinforces the position that regular aerobic, and possibly also resistance training, offers a powerful tool to cope with biologic aging of central nervous system functions. Nevertheless, the potential magnitude of cognition improvement or restrain of age-related cognition deterioration and the quantity of physical activity required to induce meaningful responses remain to be clarified.

Keywords

Brain aging • Exercise • Central noradrenergic system • Cognition • Hypothalamic-pituitary-adrenal axis • Respiratory exercise

D. Szalewska
Rehabilitation Medicine, Medical University of Gdansk, Gdansk, Poland

M. Radkowski
Department of Immunopathology of Infectious and Parasitic Diseases, Medical University of Warsaw, Warsaw, Poland

U. Demkow
Department of Laboratory Diagnostics and Clinical Immunology of Developmental Age, Medical University of Warsaw, Warsaw, Poland

P.J. Winklewski (✉)
Department of Human Physiology, Medical University of Gdansk, 15 Tuwima Street, 80-210 Gdansk, Poland

Department of Clinical Sciences, Institute of Health Sciences, Pomeranian University of Slupsk, Slupsk, Poland
e-mail: pawelwinklewski@wp.pl

1 Introduction

It is well established that regular physical activity is beneficial for both physical and mental health (Jackson et al. 2016; Kennedy et al. 2016). Consequently it is believed that regular exercise can counter typical brain aging effects. A nearly 40% decline in the number of spinal cord axons and a 10% decline in nerve conduction velocity reflects the cumulative effects of aging on central nervous system function in the course of human life. As reflexes, such as the knee-jerk, do not involve processing in the brain, aging affects them less than voluntary responses involving both reaction and movement (McArdle 2001). Many studies show that regular exercise has beneficial effects on brain health in clinical and non-clinical populations, but adherence to public health exercise guidelines is still poor (Clarkson 1978). A physically active lifestyle affects neuromuscular functions positively at any age to slow the age-related decline in cognitive performance associated with speed of information processing (Van Boxtel et al. 1997). Stimulating structural and functional adaptation that improves performance in specific tasks is the major objective of exercise training. The relationship between increased physical activity and reduced risk of Alzheimer Disease (AD) as well as increased physical activity and cognitive health have been shown in a number of large, epidemiological studies (Burley et al. 2016; Buchman et al. 2012). These adaptations require adherence to the prescribed program, with attention focused on the frequency and length of workouts, type of training, speed, intensity, duration and repetition of the activity, rest intervals and the appropriate competition.

In this review we focus on particular regimes of exercise. Physical activity can be divided, on the basis of duration of all-out exercise and the corresponding intracellular energy pathways, into four types: aerobic endurance (electron-transport oxidative phosphorylation); anaerobic power-endurance; sustained power; and strength-power (McArdle 2001). Other kinds of exercises are breathing exercises or balance training. From a psychophysiological perspective, exercise can be considered a voluntary and controllable stressor with a distinct temporal profile. Exercise activates central systems characteristic for stress response, while at the same time it may reduce the reactivity to other stressors (Stranahan et al. 2008). For example, it has been shown in rodents that 6 weeks of voluntary wheel running reduces the hypothalamic–pituitary–adrenal (HPA) axis response to other low-intensity stressors (Campeau et al. 2010).

In the first part of the article, central physiological mechanisms activated by exercise are presented with particular emphasis on the brain noradrenergic system and HPA based mostly on animal data. In the second part, the exercise strategies that optimize neuroprotection and prevent cognitive decline and neurodegenerative disease in older age are explored. The aim of this review is to find effective exercise intervention to blunt or delay the age-related development of cognitive impairment and to summarize exercise interventions that have the potential for improving brain health.

2 Central Physiological Mechanisms Activated by Exercise

Cooper (1973) were the first to link the increased peripheral concentration of norepinephrine (NE) during exercise with cognition. Those authors hypothesized that NE, crossing the blood-brain barrier, might affect the reticular formation with subsequent cognition benefit. They also were the first to point out that a moderate NE increase during medium-intensity exercise may have a positive effect on cognition, while a strong NE increase during heavy exercise exerts an opposite effect. It is currently believed that NE actually does not need to cross the blood-brain barrier. Peripherally circulating epinephrine and NE activate β-adrenoceptors of the vagus nerve (McGaugh et al. 1996). Synaptic communication between the vagal nerve and the nucleus tractus solitarius (NTS) is mediated by the excitatory neurotransmitter glutamate. As a

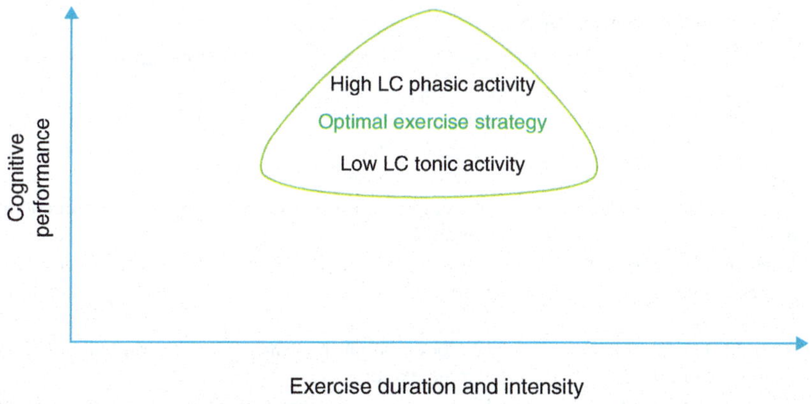

Fig. 1 Locus coeruleus (LC) phasic activity is associated with good performance in tasks requiring selective attention, while LC tonic activity with poor performance in such tasks (Rajkowski et al. 1994, Aston-Jones et al. 1998). The optimal exercise strategy, allowing for efficient non-pharmacological fine tuning of high LC phasic and low LC tonic activity, could help maintain a high signal-to-noise ratio in the central noradrenergic circuitry, resulting in improved task-oriented cognitive performance

consequence, noradrenergic cells in the NTS stimulate the locus coeruleus (LC) to synthetize and release NE (Miyashita and Williams 2006). Actually, signaling from stretch receptors, in the heart and lungs, is also relayed to the NTS via the vagal nerve (Mravec 2006; Moor et al. 2005; Berthoud and Neuhuber 2000). The glossopharyngeal and vagus nerves also relay signals from baroreceptors to the NTS (Chiba and Kato 1978; Chiba and Doba 1976). As heart rate (HR), blood pressure, and tidal breathing increase when exercise is initiated, the central NE release is liable to precede the peripheral elevation of catecholamine content (McMorris 2016). Obviously, various exercise regimes may differ regarding the elements outlined above, leading to a variability in central NE release.

The LC, a major noradrenergic nucleus in the brain, is increasingly recognized as a key element in the regulation of arousal and autonomic activity, including cognitive functions (Atzori et al. 2016; Chandler 2016; Feinstein et al. 2016; Braun et al. 2014; Samuels and Szabadi 2008a, b). During the wake state, LC displays two firing patterns, tonic and phasic, following an inverse-U shape function with respect to each other. LC's tonic firing is associated with arousal and, at the high end, limbic activation. Its phasic firing is linked to the decision-making and attention. A low level of tonic firing does not evoke a consistent behavioural response, intermediate tonic levels are associated with optimal phasic firing, and thus optimal attention and decision-making, and high tonic levels again yield low phasic response. The LC does not show any substantial activity during sleep (Aston-Jones and Waterhouse 2016; Aston-Jones and Cohen 2005; Usher et al. 1999; Aston-Jones et al. 1998; Rajkowski et al. 1994). Thus, optimal stimulation of LC function might be a crucial element to successfully modulate cognitive function with exercise (O'Donnel et al. 2012) (Fig. 1).

Neuroinflammation is an increasingly appreciated feature of many pathological conditions in the brain (Ransohoff 2016; Winklewski et al. 2015, 2016). NE suppresses the inflammatory gene transcription and augments neurotrophic factor production. The chronic activation of microglia can lead to augmented inflammatory cytokine production and subsequently chronic neuroinflammation, a recognized risk factor for various forms of dementia (Campbell 2004; Griffin et al. 1998; Benveniste et al. 1995). The anti-inflammatory

Table 1 Exercise intervention characteristics

Study	Population	Exercise strategy	Effects
Baker et al. (2010a, b)	57–83 years old; cognitively normal	6 months' high-intensity aerobic exercise	Improvement of executive functions and insulin sensitivity
Muscari et al. (2010)	57–83 years old; cognitively normal	12 months' high-intensity aerobic exercise	Increase in hippocampal volumes
Kemoun et al. (2010)	81.8 ± 1.8 years old; with dementia	15 weeks' high-intensity aerobic exercise	Improvement in cognition
Makizako et al. (2012)	Mean age 76 years; with amnestic mild cognitive impairment	6 months' multicomponent exercise program	No effects on dual-task performance assessed by reaction time
Winchester et al. (2013)	81 ± 11.3 years old; cognitively normal	12 months' high-intensity aerobic exercise	Improvement in cognition
Chapman et al. (2013)	74.5 ± 4.5 years old; cognitively normal	12 weeks' aerobic exercise	Improvement in cognition and cerebral blood flow, especially in anterior cingulate region
Fallah et al. (2013)	65–75 years old women; cognitively normal	Strength power exercise training	Improvement in cognition
Singh et al. (2014)	70.1 ± 6.7 years old; with mild cognitive impairment	6 months' high-intensity progressive resistance training	Improvement in global cognitive and executive functions
ten Brinke et al. (2015)	70–80 years old women; with mild cognitive impairment	26 weeks' aerobic exercises	Improvement in hippocampal volume
Ferreira et al. (2015)	Elderly individuals; cognitively normal	6 months' respiratory training	Improvement in cognition in relation to abstraction and mental flexibility
Bossers et al. (2015)	85.5 ± 5.1 years old in a psycho-geriatric nursing home; with dementia	9 weeks' combination of aerobic and strength training	Improvement in cognition
Chapman et al. (2016)	56–75 years old;, cognitively normal	12-weeks' aerobic training at heart rate obtained at 50–75% VO_{2max}	Improvement in cerebral blood flow and memory

action of NE is exerted via β_2-adrenergic receptors present on microglia and astrocytes (Hetier et al. 1991; Frohman et al. 1998). Other anti-inflammatory effects of NE have also been reported, including the suppression of inducible nitric oxide synthase, interleukin-1b, tumor necrosis factor α, and intercellular adhesion molecule-1. Beside anti-inflammatory effects, NE enhances brain-derived neurotrophic factor production, which plays an important role in neuronal survival, neuroplasticity, and neurogenesis. The enhancement of brain-derived neurotrophic factor is mediated by β_1/β_2, and α_2-adrenergic receptors and shares similar cellular pathways with anti-inflammatory NE action (Juric et al. 2008; Binder and Scharfman 2004; Zafra et al. 1992).

The stress-induced activation of the HPA axis and the release of glucocorticoid hormones, cortisol in humans and corticosterone in rodents, affect learning and memory via modulation of mineralocorticoid and glucocorticoid receptors (Roozendaal et al. 2009; Sandi Pinelo-Nava 2007; Joels et al. 2006). Stimulation of the mineralocorticoid receptors is beneficial to a wide array of cognitive and affective functions. Nevertheless, the precise cognitive process that mediates such a broad range of positive effects has yet to be elucidated. Importantly, apart from beneficial actions, cortisol augments the reconsolidation of fear memory, which underscores the detrimental side of central glucocorticoid receptors stimulation (Wolf et al. 2016). The influence of the stress response on central physiological mechanisms highly depends on predictability and controllability. A stressor is 'controllable' if it can be initiated or terminated by the subject. Likewise, a stressor is 'predictable' if it occurs at temporary intervals that are regularly spaced. If the stressor is controllable and predictable, the threat-recognition regions, including the amygdala, change the involvement pattern in the stress response. Voluntary running exercise diminishes the density of activated noradrenergic cells in the basolateral amygdala in rodents, and increases the density of such cells in the central nucleus of the amygdala (Burghardt et al. 2006). Apart from altering basal patterns of activation, a sudden inability to start wheel running activates noradrenergic cells in the amygdala (Rhodes et al. 2003). When translating the animal data to humans it should be kept in mind that running is a rewarding, voluntary form of stress for rodents. Therefore, it may actually include the activation of the reward system of these animals (Stranahan et al. 2008). Moreover, for obvious reasons human studies are performed in a far less controlled environment than in laboratory animals.

3 Exercise Strategies to Optimize Neuroprotection in Humans

We searched 'PubMed' from January 1996 to September 2016 for any clinical trials investigating the influence of different kinds of exercise on cognitive decline. Twelve randomized and non-randomized trials, which were focused preferably on the effects of physical exercise, discounting diet or other lifestyle changes, were included in the present review. The intervention characteristics are shown in Table 1. A review from the Cochrane Database was used to summarize the biological effects of exercise (Forbes et al. 2015). None of these studies have focused on the assessment of the brain noradrenergic system. Such an assessment has become recently possible in humans, taking the pupil diameter as a non-invasive surrogated of LC activity (Costa and Rudebeck 2016; Joshi et al. 2016).

3.1 Aerobic Endurance – Longer-Term High-Intensity Aerobic Exercise

Baker et al. (2010a) have randomized 34 subjects (57–83 years of age) with glucose intolerance and normal cognitive status to aerobic exercise or stretching control group. Exercise in both groups was carried out 4 day/week for 45–60 min per session for 6 months. The first eight sessions were supervised by a fitness trainer and then one session/week/participant was performed at home. Exercise duration and intensity were titrated up over the first 6 weeks, until participants in the aerobic group were at 75–85% of heart rate (HR) reserve using a treadmill, stationary bicycle, or elliptical trainer. Participants in the control group carried out a prescribed routine of balance and stretching exercises, maintaining HR at or below 50% HR reserve. Aerobic exercise improved executive function measured by Trails B, Task Switching, Stroop, Self-ordered Pointing Test and verbal fluency, cardiorespiratory fitness, and insulin sensitivity. In another trial, supervised by the same authors, the effects of aerobic exercise on cognition and other biomarkers associated with Alzheimer's disease (AD) pathology were examined in 33 adults aged from 55 to 85 years, with amnestic mild cognitive impairment (Baker et al. 2010b). The design of the exercise intervention was like that of the trial above described. Six months of high-intensity aerobic exercise had sex-specific effects on cognition, glucose metabolism, HPA axis, and trophic activity despite comparable gains in cardiorespiratory fitness and body fat reduction. In women, aerobic exercise improved performance on multiple tests of executive function, increased glucose disposal during the metabolic clamp, and reduced fasting plasma levels of insulin-like growth factor-1 and had a favorable effect only on Trails B performance.

3.2 Aerobic Endurance – Longer-Term Moderate-Intensity Aerobic Exercise

Muscari et al. (2010) have performed a 12-month supervised endurance exercise training 3 h a week in 120 healthy subjects aged 65–74 years. The authors have concluded that a 12-month reduces the progression of age-related cognitive decline in healthy older adults as assessed by the Mini-Mental State Examination.

3.3 Aerobic Endurance – Longer-Term Low-Intensity Training: Walking

A brain health study (BHS), a sub-study within the larger Baltimore Experience Corps Trial (BECT), a sex-stratified randomized controlled trial was designed to evaluate the health benefits for older adults in 123 participants aged over 60 years. Walking activity was measured using a step monitor. The authors found that greater daily walking activity was associated with larger hippocampal volumes among older women, but no men (Varma et al. 2015). Very interesting data on the of exercise in a group of 104 older adults with AD aged 81 ± 6.5 years were presented by Winchester et al. (2013). After 1 year of follow-up the authors have concluded that a sedentary lifestyle correlates with a decline in cognitive function, a loss of vigor and increases in the feelings of anger, confusion, depression, and fatigue. Some level of physical activity, especially walking, was beneficial to cognitive function in subjects with mild-to-moderate AD. Similar results concerning the effects of a 24-month walking were obtained in the program on 21 elderly nursing home residents (84 ± 5 years old) with AD. The training group showed significant improvement in the Mini-Mental State Examination, which may indicate that this specific walking program may stabilize the progressive cognitive dysfunctions in such patients (Venturelli et al. 2011). Kemoun et al. (2010) have obtained similar results using the Rapid Evaluation of Cognitive Functions Test in 31 subjects aged 81.8 ± 5.3 years suffering from dementia after 15 weeks of rehabilitation, consisting of walking, equilibrium, and endurance exercise in 1-h sessions per week. Ten Brinke et al. (2015) conducted a 26-week single-blinded, randomized, twice-weekly program, in which 39 women with probable mild cognitive impairment, aged 70–80 years, were randomized into three groups: 1) aerobic exercises (AT); 2) resistance training (RT); or 3) balance and tone training (BAT). The classes were 60 min in duration (10-min warm-up, 40 min of core content, and 10-min cool down). The authors found that both left and right hippocampal volumes assessed by magnetic resonance imaging (MRI) enlarged significantly over the course of the intervention in the aerobic training group compared with the BAT group. In contrast to aerobic training, resistance training did not have a significant effect on hippocampal volume. The role of increased hippocampal volume in response to aerobic training remains unsettled as the increased hippocampal volume was independently associated with reduced verbal memory and learning performance.

3.4 Aerobic Endurance – Shorter-Term Aerobic Exercise

Chapman et al. (2016) have randomized 37 cognitively normal adults aged 74.5 ± 5.8 years into two different groups: physical training or wait-list control. The physical training program comprised of 150 min exercise per week, as recommended for sedentary adults. The training regimen lasted 12 weeks and consisted of three 60 min sessions of aerobic exercise; each session alternating between exercise bike and treadmill. The bike exercise included 5 min warm up at 43 watts, cycling for 50 min at a speed that increased HR to 50–76% of the maximum achieved on VO_2 max testing, and a 5 min cool down at 43 watts. The treadmill workout included 5 min warm up at 2 miles per hour, walking on the treadmill for 50 min at a speed that increased their heart rate to 50–75% of the maximum achieved on VO_2max testing, and a 5 min cool down at 2 miles per hour. The authors have concluded that this kind of exercise improves cardiovascular fitness, cognition, and regional cerebral blood flow assessed in MRI, particularly noted in the anterior cingulate region. In an interesting trial performed by the same research group the effects of two training protocols, cognitive training vs. physical training, on cognition and brain function were evaluated in adults aged 56–75 (Chapman et al. 2013). The exercise training program was the same as that in the study above outlined. The results were that cognitive training improved executive function whereas physical training enhanced memory. Subjects from the physical training group showed higher cerebral blood flow in the hippocampal region, an area particularly vulnerable to aging and dementia.

3.5 Aerobic Exercise Vs. Respiratory Training

Interesting data have been provided by the group of Ferreira et al. (2015), who performed a controlled study in 68 elderly individuals randomized into three groups: aerobic exercise group (the 'walking' group); respiratory training group (the 'breathing' group); and the social interaction group (the control group). The first group followed a program of supervised walking that lasted 6 months, three times a week, with a duration of 40–50 min each session. The intensity was moderate and was limited to a range of 60–80% of HR reserve. The second group focused specifically on respiratory exercise with a duration and intensity like in the 'walking group', which comprised: (1) stretching exercises for muscles of the trunk, the neck and the upper limbs; (2) seven breathing exercises; and (3) inspiratory muscle training with a RESPIRON. The authors found that only the breathing group improved cognitive function in relation to abstraction and mental flexibility, understood as the number of perseverative errors in the test.

3.6 Sustained and Strength Power – Resistance Training

In the randomized, double-blind, double-sham controlled trial Study of Mental and Resistance Training (SMART) in one hundred adults with mild cognitive impairment and a mean age of 70.1 years, 6-months' high-intensity progressive resistance training 2–3 times a week significantly improved global cognitive function assessed by Alzheimer's Disease Assessment Scale-cognitive subscale at 6 months and executive function measured by Wechsler Adult Intelligence Scale Matrices across 18 months (Singh et al. 2014). Fallah et al. (2013) have investigated the effect of a targeted 60-min exercise training on the probability of an improved, maintained, or declined performance of executive functions in elderly independently living women aged 65–75, who scored >24 on the Mini-Mental State Examination and had a visual acuity of at least 20/40 without corrected lenses. Keiser-based exercises consisted of biceps curls, triceps extension, seated row, latissimus dorsi pull-downs, leg press, hamstring curls, and calf raises. The intensity of the training stimulus was at a work range of 6–8 repetitions (two sets) and was subsequently increased. Other key strength exercises included mini-squats, mini-lunges, and lunge walks. The balance and tone training program consisted of stretching exercises, a range of motion exercises, balance exercises and functional sand relaxation technique. The authors found that the resistant training group had a significantly higher performance on the Stroop Test compared with balance and tone training. For shifting and working memory there were no significant differences between the groups.

3.7 Combination of Aerobic and Resistance Training and Multicomponent Exercise Training

Bossers et al. (2015) in a group of 109 patients with dementia from a psycho-geriatric nursing home (aged 85.5 ± 5.1 years) have shown, compared with non-exercising control subjects, that a combination of aerobic and strength training is more effective than aerobic-only training in slowing cognitive and motor decline. It was 9-week, parallel, three-group, single-blinded, randomized controlled trial with a follow-up assessment at week 18. In a randomized controlled trial, carried out in 50 subjects of the mean age of 76 years, with amnestic mild cognitive impairment, a 6-month multicomponent exercise program improved only maximal walking speed, with no effects on dual-task performances assessed by the reaction time (Makizako et al. 2012). An interesting randomized, controlled trial concerning the promising effects of dance movement intervention and exercise on elderly with early dementia has been proposed by Ho et al. (2015). Another randomized study proposed by Martinez-Velilla et al. (2015) for hospitalized patients aged 75 years and older, who are taking part in an exercise program composed of supervised progressive resistance training, balance training, and walking for 5–7 consecutive days, is currently underway. Primary outcome measures are changes in functional status after the intervention, assessed with the Mini-Mental State Examination, GDS Yesavage, and Trail Making Test. Biological plausibility of the effects of exercise intervention on cognitive outcomes in relation to the kind of exercise has been clearly summarized in the appendix of the document from the Cochrane Library (Forbes et al. 2015).

4 Summary

4.1 Aerobic Exercise

Aerobic exercise diminishes oxidative stress, reduces age-related microglial priming and activation, and augments microglia cells volume, which is protective and could aid neuroprotection. Further, aerobic exercise increases production improves signaling of such growth factors as brain-derived neurotrophic factor, insulin growth factor-1, and vascular endothelial growth factor, which all are associated with neurogenesis and angiogenesis. Finally, aerobic physical activity enhances brain insulin signaling, which has a vasodilating influence that may improve cerebral blood flow. Such changes lead to the augmentation of the anterior hippocampal and medial temporal lobe volumes, increase hippocampal capillary density and branching, and diminish extracellular amyloid-β deposition in the brain.

4.2 Resistance Exercise

Resistance exercise is likely to promote maintenance of cerebral perfusion, increases the concentration of brain-derived neurotrophic factor, insulin growth factor-1, and vascular endothelial growth factor, all associated with neurogenesis, and improves brain insulin signaling.

5 Conclusions

The evidence from animal studies shows that the central noradrenergic system and the hypothalamic-pituitary-adrenal axis are modulated by exercise. However, the difference in the activation pattern of either by voluntary (e.g., exercise) and involuntary (e.g., fear or anxiety) stressors remains obscure. Recent advances in the field and a better understanding of central physiological mechanisms governing stress response clearly open new avenues for therapeutic translation.

The accumulating evidence reinforces the position that regular aerobic and, with less evidence of effectiveness, resistance training offer a powerful tool to cope with biologic aging of selected central nervous system functions. Nevertheless, the potential magnitude of cognition improvement, or slowdown of cognition deterioration, and the quantity of physical activity required to induce meaningful responses remain to be clarified. A large variety of cognitive tests used to assess the response to various exercise regimes make direct comparisons among particular experimental settings difficult. Further studies and randomized clinical trials are needed to establish guidelines for using exercise as a therapeutic option.

Conflicts of Interest The authors declare no conflicts of interest in relation to this article.

References

Aston-Jones G, Cohen JD (2005) An integrative theory of locus coeruleus-norepinephrine function: adaptive gain and optimal performance. Annu Rev Neurosci 28:403–450

Aston-Jones G, Waterhouse B (2016) Locus coeruleus: from global projection system to adaptive regulation of behavior. Brain Res 1645:75–78

Aston-Jones G, Rajkowski J, Ivanova S, Usher M, Cohen J (1998) Neuromodulation and cognitive performance: recent studies of noradrenergic locus coeruleus neurons in behaving monkeys. Adv Pharmacol 42:755–759

Atzori M, Cuevas-Olguin R, Esquivel-Rendon E, Garcia-Oscos F, Salgado-Delgado RC, Saderi N, Miranda-Morales M, Treviño M, Pineda JC, Salgado H (2016) Locus coeruleus norepinephrine release: a central regulator of CNS spatio-temporal activation? Front Synaptic Neurosci 8:25

Baker LD, Frank LL, Foster-Schubert K, Green PS, Wilkinson CW, McTiernan A, Cholerton BA, Plymate SR, Fishel MA, Watson GS, Duncan GE, Mehta PD, Craft S (2010a) Aerobic exercise improves cognition for older adults with glucose intolerance, a risk factor for Alzheimer's disease. J Alzheimers Dis 22:569–579

Baker LD, Frank L, Foster-Schubert K, Green PS, Wilkinson CW, McTiernan A, Plymate SR, Fishel MA, Watson GS, Cholerton BA, Duncan GE, Mehta PD, Craft S (2010b) Effects of aerobic exercise on mild cognitive impairment: a controlled trial. Arch Neurol 67:71–79

Benveniste EN, Huneycutt BS, Shrikant P, Ballestas ME (1995) Second messenger systems in the regulation of cytokines and adhesion molecules in the central nervous system. Brain Behav Immun 9:304–314

Berthoud HR, Neuhuber WL (2000) Functional and chemical anatomy of the afferent vagal system. Auton Neurosci 85:1–17

Binder DK, Scharfman HE (2004) Brain-derived neurotrophic factor. Growth Factors 22:123–131

Bossers WJ, van der Woude LH, Boersma F, Hortobágyi T, Scherder EJ, van Heuvelen MJ (2015) A 9-week aerobic and strength training program improves cognitive and motor function in patients with dementia: a randomized, controlled trial. Am J Geriatr Psychiatry 23:1106–1611

Braun D, Madrigal JL, Feinstein DL (2014) Noradrenergic regulation of glial activation: molecular mechanisms and therapeutic implications. Curr Neuropharmacol 12:342–352

Buchman AS, Boyle PA, Yu L, Shah RC, Wilson RS, Bennett DA (2012) Total daily physical activity and the risk of AD and cognitive decline in older adults. Neurology 78(17):1323–1329

Burghardt PR, Pasumarthi RK, Wilson MA, Fadel J (2006) Alterations in fear conditioning and amygdalar activation following chronic wheel running in rats. Pharmacol Biochem Behav 84:306–312

Burley CV, Bailey DM, Marley CJ, Lucas SJ (2016) Brain train to combat brain drain; focus on exercise strategies that optimise neuroprotection. Exp Physiol. doi:10.1113/EP085672

Campbell A (2004) Inflammation, neurodegenerative diseases, and environmental exposures. Ann N Y Acad Sci 1035:117–132

Campeau S, Nyhuis TJ, Sasse SK, Kryskow EM, Herlihy L, Masini CV, Babb JA, Greenwood BN, Fleshner M, Day HE (2010) Hypothalamic pituitary adrenal axis responses to low-intensity stressors are reduced after voluntary wheel running in rats. J Neuroendocrinol 22:872–888

Chandler DJ (2016) Evidence for a specialized role of the locus coeruleus noradrenergic system in cortical circuitries and behavioral operations. Brain Res 1641:197–206

Chapman SN, Aslan S, Spence JS, Defina LF, Keebler MW, Didehbani N, Lu H (2013) Shorter term aerobic exercise improves brain, cognition, and cardiovascular fitness in aging. Front Aging Neurosci 5:1–9

Chapman SN, Aslan S, Spence JS, Keebler MW, DeFina LF, Didehbani N, Perez AM, Lu H, D'Esposito M (2016) Distinct brain and behavioral benefits from cognitive vs. physical training: a randomized trial in aging adults. Front Aging Neurosci 10:1–15

Chiba T, Doba N (1976) Catecholaminergic axo-axonic synapses in the nucleus of the tractus solitarius (pars commissuralis) of the cat: possible relation to presynaptic regulation of baroreceptor reflexes. Brain Res 102:255–665

Chiba T, Kato M (1978) Synaptic structures and quantification of catecholaminergic axons in the nucleus tractus solitarius of the rat: possible modulatory roles of catecholamines in baroreceptor reflexes. Brain Res 151:323–338

Clarkson PM (1978) The relationship of age and level of physical activity with the fractionated components of patellar reflex time. J Gerontol 3:650

Cooper CJ (1973) Anatomical and physiological mechanisms of arousal, with special reference to the effects of exercise. Ergonomics 16:601–609

Costa VD, Rudebeck PH (2016) More than meets the eye: the relationship between pupil size and locus coeruleus activity. Neuron 89:8–10

Fallah N, Hsu CL, Bolandzadeh N, Davis J, Beattie BL, Graf P, Liu-Ambrose TA (2013) Multi-state model of cognitive dynamics in relation to resistance training:

the contribution of baseline function. Ann Epidemiol 23:463–468

Feinstein DL, Kalinin S, Braun D (2016) Causes, consequences, and cures for neuroinflammation mediated via the locus coeruleus: noradrenergic signaling system. J Neurochem. doi:10.1111/jnc.13447

Ferreira L, Tanaka K, Santos-Galduróz RF, Galduróz JC (2015) Respiratory training as a strategy to prevent cognitive decline in aging: a randomized controlled trial. Clin Interv Aging 10:593–603

Forbes SC, Forbes D, Forbes S, Blake CM, Chong LY, Thiessen EJ, Little JP, Rutjes AWS (2015) Exercise interventions for preventing dementia or delaying cognitive decline in people with mild cognitive impairment (Protocol). Cochrane Database Syst Rev 5, art. no CD011706

Frohman EM, Vayuvegula B, Gupta S, van den Noort S (1998) Norepinephrine inhibits gamma-interferon-induced major histocompatibility class II (Ia) antigen expression on cultured astrocytes via beta-2-adrenergic signal transduction mechanisms. Proc Natl Acad Sci U S A 85:1292–1296

Griffin WS, Sheng JG, Royston MC, Gentleman SM, McKenzie JE, Graham DI, Roberts GQ, Mark RE (1998) Glial-neuronal interactions in Alzheimer's disease: the potential role of a 'cytokine cycle' in disease progression. Brain Pathol 8:65–72

Hetier E, Ayala J, Bousseau A, Prochiantz A (1991) Modulation of interleukin-1 and tumor necrosis factor expression by beta-adrenergic agonists in mouse ameboid microglial cells. Exp Brain Res 86:407–413

Ho RT, Cheung JK, Chan WC, Cheung IK, Lam LC (2015) A 3-arm randomized controlled trial on the effects of dance movement intervention and exercises on elderly with early dementia. BMC Geriatr 15:1

Jackson PA, Pialoux V, Corbett D, Drogos L, Erickson KI, Eskes GA, Poulin MJ (2016) Promoting brain health through exercise and diet in older adults: a physiological perspective. J Physiol 594:4485–4498

Joels M, Pu Z, Wiegert O, Oitzl MS, Krugers HJ (2006) Learning under stress: how does it work? Trends Cogn Sci 10:152–158

Joshi S, Li Y, Kalwani RM, Gold JI (2016) Relationships between pupil diameter and neuronal activity in the locus coeruleus, colliculi, and cingulate cortex. Neuron 89:221–234

Juric DM, Loncar D, Carman-Krzan M (2008) Noradrenergic stimulation of BDNF synthesis in astrocytes: mediation via alpha1- and beta1/beta2-adrenergic receptors. Neurochem Int 52:297–306

Kemoun G, Thibaud M, Roumagne N, Carette P, Albinet C, Toussaint L, Paccalin M, Dugué B (2010) Effects of a physical training programme on cognitive function wand walking efficiency in elderly persons with dementia. Dement Geriatr Cogn Disord 29:109–114

Kennedy G, Hardman RJ, Macpherson H, Scholey AB, Pipingas A (2016) How does exercise reduce the rate of age-associated cognitive decline? A review of potential mechanisms. J Alzheimers 55(1):1–18

Makizako H, Doi T, Shimada H, Yoshida D, Tsutsumimoto K, Uemura K, Suzuki T (2012) Does a multicomponent exercise program improve dual-task performance in amnestic mild cognitive impairment? A randomized controlled trial. Aging Clin Exp Res 24:640–646

Martinez-Velilla N, Casas-Herrero A, Zambom-Ferraresi-F, Suarez N, Alonso-Renedo J, Contin KC, de Astesau ML, Echeverria NF, Lazaro MG, Izquierdo M (2015) Functional and cognitive impairment prevention through early physical activity for geriatric hospitalized patients: study protocol for a randomized controlled trial. BMC Geriatr 15:112

McArdle WD (2001) Exercise physiology: energy, nutrition and human performance. Lippincott Williams and Wilkins, Philadelphia

McGaugh JL, Cahill L, Roozendaal B (1996) Involvement of the amygdala in memory storage: interaction with other brain systems. Proc Natl Acad Sci U S A 93:13508–13514

McMorris T (2016) Developing the catecholamines hypothesis for the acute exercise-cognition interaction in humans: lessons from animal studies. Physiol Behav 165:291–299

Miyashita T, Williams CL (2006) Epinephrine administration increases neural impulses propagated along the vagus nerve: role of peripheral beta-adrenergic receptors. Neurobiol Learn Mem 85:116–124

Moor T, Mundorff L, Bohringer A, Philippsen C, Langewitz W, Reino ST, Schachinger H (2005) Evidence that baroreflex feedback influences long-termincidental visual memory in men. Neurobiol Learn Mem 84:168–174

Mravec B (2006) Possible involvement of the vagus nerve in monitoring plasma catecholamine levels. Neurobiol Learn Mem 86:353–355

Muscari A, Giannoni C, Pierpaoli L, Berzigotti A, Maietta P, Foschi E, Ravaioli C, Poggiopollini G, Bianchi G, Magalotti D, Tentoni C, Zoli M (2010) Chronic endurance exercise training prevents aging-related cognitive decline in healthy older adults: a randomized controlled trial. Int J Geriatr Psychiatry 25:1055–1064

O'Donnell J, Zeppenfeld D, McConnell E, Pena S, Nedergaard M (2012) Norepinephrine: a neuromodulator that boosts the function of multiple cell types to optimize CNS performance. Neurochem Res 37:2496–2512

Rajkowski J, Kubiak P, Aston-Jones G (1994) Locus coeruleus activity in monkey: phasic and tonic changes are associated with altered vigilance. Brain Res Bull 35:607–616

Ransohoff RM (2016) How neuroinflammation contributes to neurodegeneration. Science 353:777–783

Rhodes JS, Garland T Jr, Gammie SC (2003) Patterns of brain activity associated with variation in voluntary wheel-running behavior. Behav Neurosci 117:1243–1256

Roozendaal B, McEwen BS, Chattarji S (2009) Stress, memory and the amygdala. Nat Rev Neurosci 10:423–433

Samuels ER, Szabadi E (2008a) Functional neuroanatomy of the noradrenergic locus coeruleus: its roles in the regulation of arousal and autonomic function part I: principles of functional organisation. Curr Neuropharmacol 6:235–253

Samuels ER, Szabadi E (2008b) Functional neuroanatomy of the noradrenergic locus coeruleus: its roles in the regulation of arousal and autonomic function part II: physiological and pharmacological manipulations and pathological alterations of locus coeruleus activity in humans. Curr Neuropharmacol 6:254–285

Sandi C, Pinelo-Nava MT (2007) Stress and memory: behavioral effects and neurobiological mechanisms. Neural Plast 2007:78970

Singh MAF, Gates N, Saigal N, Wilson GC, Meiklejohn J, Brodaty H, Wen W, Baune BT, Suo C, Baker MK, Foroughi N, Wang Y, Sachdey PS, Valenzuela M, The Study of Mental and Resistance Training (SMART) (2014) Study – resistance training and/or cognitive training in mild cognitive impairment: a randomized, double-blind, double-sham controlled trial. JAMDA 15:873–880

Stranahan AM, Lee K, Mattson MP (2008) Central mechanisms of HPA axis regulation by voluntary exercise. Neruomol Med 10:118–127

ten Brinke LF, Bolandzadeh N, Nagamatsu LS, Hsu CL, Davis JC, Miran-Khan K, Liu-Ambrose T (2015) Aerobic exercise increases hippocampal volume in older women with probable mild cognitive impairment: a 6-month randomized controlled trial. Br J Sports Med 49:248–254

Usher M, Cohen JD, Servan-Schreiber D, Rajkowski J, Aston-Jones G (1999) The role of locus coeruleus in the regulation of cognitive performance. Science 283:549–554

Van Boxtel MPJ, Paas FG, Houx PJ, Adam JJ, Teeken JC, Jolles J (1997) Aerobic capacity and cognitive performance in a cross-sectional aging study. Med Sci Sports Exerc 29:1357

Varma VR, Chuang Y, Harris GC, Tan EJ, Carlson MC (2015) Low-intensity daily walking activity is associated with hippocampal volume in older adults. Hippocampus 25:605–615

Venturelli M, Scarsini R, Schena F (2011) Six-month walking program changes cognitive and ADL performance in patients with Alzheimer. Am J Alzheimers Dis Other Demen 26:381–389

Winchester J, Dick MB, Gillen D, Reed B, Miller B, Tinklenberg J, Mungas D, Chui H, Galasko D, Hewett L, Cotman CW (2013) Walking stabilizes cognitive functioning in Alzheimer's disease across one year. Arch Gerontol Geriatr 56:96–103

Winklewski PJ, Radkowski M, Wszedybyl-Winklewska-M, Demkow U (2015) Brain inflammation and hypertension: the chicken or the egg? J Neuroinflammation 12:85

Winklewski PJ, Radkowski M, Demkow U (2016) Neuroinflammatory mechanisms of hypertension: potential therapeutic implications. Curr Opin Nephrol Hypertens 25:410–416

Wolf OT, Atsak P, de Quervain DJ, Roozendaal B, Wingenfeld K (2016) Stress and memory: a selective review on recent developments in the understanding of stress hormone effects on memory and their clinical relevance. J Neuroendocrinol. doi:10.1111/jne.12353

Zafra F, Lindholm D, Castren E, Hartikka J, Thoenen H (1992) Regulation of brain-derived neurotrophic factor and nerve growth factor mRNA in primary cultures of hippocampal neurons and astrocytes. J Neurosci 12:4793–4799

Advs Exp. Medicine, Biology - Neuroscience and Respiration (2017) 32: 81–89
DOI 10.1007/5584_2017_20

Published online: 17 May 2017

Evaluation of Immune Indices and Serum Vitamin D Content in Children with Atopic Dermatitis

Agnieszka Lipińska-Opałka, Agata Wawrzyniak, Sławomir Lewicki, Robert Zdanowski, and Bolesław Kalicki

Abstract

The influence of vitamin D on allergic diseases, including atopic dermatitis, is linked to the presence of vitamin D nuclear receptors in immune cells. The present study seeks to determine the possible relationship between serum vitamin D content and immune indices in children with atopic dermatitis. The study was conducted in 19 children with atopic dermatitis. The control consisted of 17 age-matched healthy children. A single significant finding was a distinctly lower number of serum regulatory T cells in atopic dermatitis compared with controls ($p < 0.00001$). There were no appreciable differences between the two groups concerning the immunological indices such as the phenotypes: CD3, CD4, CD8, CD4/CD8, CD19, CD16/56, natural killer T cells, and anti-CD3 human leukocyte antigen – antigen D related cell surface receptor (HLA-DR3), or the percentage of lymphocytes, eosinophils, and the IgE level. We also revealed an inverse association between the serum vitamin D and the percentage of CD8+ cells ($p < 0.05$; $r = 0.62$) in atopic dermatitis. In conclusion, the results point to a regulatory role of T cells in the pathogenesis of atopic dermatitis, but fail to substantiate the influence of vitamin D on the course of the disease.

Keywords

Allergy • Atopic dermatitis • Children • Regulatory T cells • Vitamin D

A. Lipińska-Opałka (✉), A. Wawrzyniak, and B. Kalicki
Pediatric, Nephrology and Allergology Clinic, Military Institute of Medicine, Warsaw, Poland
e-mail: lipinska.ag@gmail.com

S. Lewicki and R. Zdanowski
Department of Regenerative Medicine and Cell Biology, Military Institute of Hygiene and Epidemiology, Warsaw, Poland

1 Introduction

Atopic dermatitis is a chronic inflammatory skin disease related with epidermis and dermis dysfunction. It is characterized by a chronic and recurrent course and distinctive morphology and location of lesions. The concept of atopic dermatitis was introduced in 1933 by Wise and Sulzberger (Reitamo et al. 2008), while the diagnostic criteria were proposed in 1980 by Hanifin and Rajka (1980) and they are still valid (Darsow et al. 2010). The disease usually begins in early childhood; 60% of cases occur before one year of age and in up to 45% of cases the first symptoms appear before 6 months of age. The incidence of allergic diseases, including atopic dermatitis, increases in recent years, which is related to civilization development. Over the past 30 years, two- or even threefold increase in the morbidity of atopic dermatitis has been observed in industrialized countries (Bieber 2010). The incidence of the disease in the European population is estimated at 10–12% (Romański 1998), and it reaches 15–30% in pediatric patients (Bieber 2010). A higher percentage of cases is observed in urban environments, which highlights the influence of the surrounding environment on the development and course of the disease (Williams and Flohr 2006). Atopic dermatitis is often the first manifestation of allergic disease, which course has a form of so-called atopic march. This means that symptoms of atopic dermatitis can overtake the syndromes of asthma and allergic rhinitis (Kapoor et al. 2008). The main clinical symptoms of atopic dermatitis include erythematous papules, thickened skin with a tendency to lichenification, and severe pruritus.

The pathogenesis of atopic dermatitis is complex and not fully understood. The disease is a result of genetic, environmental, and immune disorders. Atopic dermatitis is a complex inflammatory process, in which T cells, macrophages, eosinophils, and immunoglobulin E (IgE) play an essential role. In the acute phase of the disease, there is an imbalance between Th1 and Th2-responses (Boguniewicz 2014; Leung 2000). Elevated Th2-expressing cells in the peripheral blood cause increases in IL-4, IL-5, and IL-13 concentration (Ong and Leung 2006). These cytokines stimulate B cells to excessive production of immunoglobulin E and enhance eosinophils proliferation. Furthermore they are capable of inhibition of the Th1 cells cytokine expression and maturation. However, in the chronic phase there is a domination of Th1-responses. Excessive activity of both pathways may be inhibited by the regulatory T lymphocytes (Treg, Fox P3), but their number and activity is markedly reduced in atopic dermatitis.

The pleiotropic effect of vitamin D in the human body is widely discussed. This vitamin also plays an important role in atopic dermatitis (Mesquita et al. 2013). Numerous studies have shown that the appropriate level of vitamin D can affect the T cells activation and the number of regulatory T cells (Vassallo and Camargo 2010). Nowadays, deficiency of vitamin D in the European population is commonly observed. This is mostly caused by a change of people's lifestyle during last decades consisting of spending more time indoors, use of UV filters, and dietic vitamin D deficiency.

The aim of the present study was to evaluate the percentage of natural T regulatory cells (nTreg, $CD4^+$, $CD25^{++}$, $FoxP3^+$, and $CD127^-$) and lymphocyte phenotypes ($CD3^+$, $CD4^+$, $CD8^+$, CD4/CD8 ratio, $CD19^+$ $CD16^+/56^+$, and $CD3^+$ anty-HLA-DR^+) in children with atopic dermatitis.

2 Methods

2.1 Patients

The study was approved by the Ethics Committees of the Military Institute of Medicine in Warsaw (permit no. 123/14) and the parents of all participants gave written informed consent. The study group consisted of 19 children with atopic dermatitis, with the median age of 3 years (IQR 2–6 years). The inclusion criteria were the age from 6 months to 15 years and the diagnosis of atopic dermatitis as based on the criteria of Hanifin and Rajka (1980). A suspicion of other skin diseases excluded patients from the study group. Atopic dermatitis was evaluated by the SCORAD scale, an index of its severity, which takes into consideration the extent and severity of lesions, and the intensity of subjective symptoms such as itching and insomnia. The control group consisted of 17 children with no history of atopy and no skin lesions. The median age in this group was 3 years (IQR 2–5 years).

2.2 Blood Lymphocyte Phenotype Determination

Blood lymphocyte phenotypes were determined as previously described (Kalicki et al. 2013). Briefly, peripheral blood samples were collected in EDTA-anticoagulated tubes. The lymphocyte immunophenotype was determined using BD Simultest™ – IMK Plus Kit (BD Biosciences, Warsaw, Poland). Blood samples (100 μL) were incubated for 20 min with appropriate antibodies and then erythrocyte lysis was performed in the dark at room temperature for 10 min (FACS Lysing Solution; BD Biosciences). Finally, cells were washed twice with 2 ml of phosphate buffered saline (PBS) and fixed in 200 μL of 1% paraformaldehyde in PBS. Then, cells were examined by flow cytometry (FACS Calibur; BD Biosciences). Additionally, percentage distribution of white blood cells (lymphocytes, monocytes, granulocytes, neutrophils, and eosinophils) was assessed using CD45 FITC and CD14 PE antibodies, and FSC/SSC

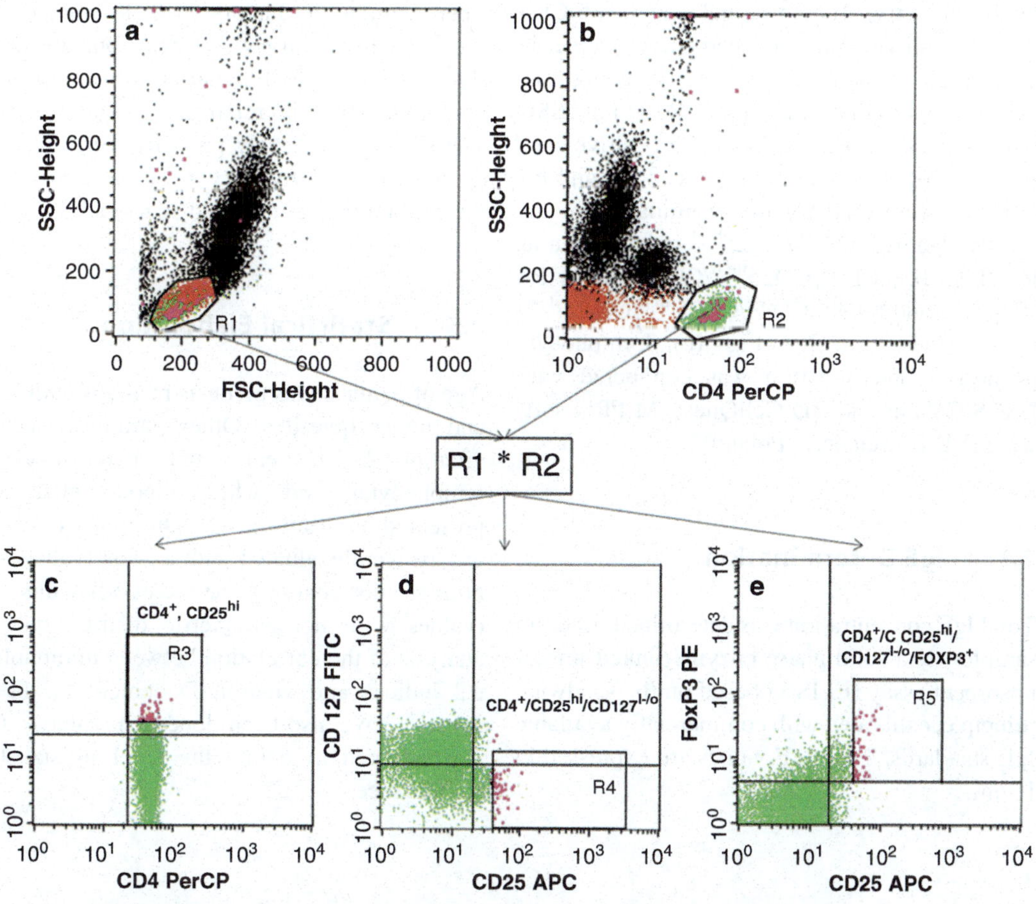

Fig. 1 Gating strategy in assessing nTreg populations (see phenotype determination for details)

determination. The results of lymphocyte phenotype assessment are presented as mean percentages of lymphocytes ± SD.

To characterize T regulatory cells, whole blood samples (100 μL) were stained with the primary antibodies CD4-PerCP, CD25-APC, and CD127-FITC (BD Biosciences, Warsaw, Poland) or with an appropriate isotype control with the addition of CD4-PerCP antibody. Blood samples were incubated for 20 min in the dark at room temperature and the erythrocyte lysis was performed (BD FACS Lysing Solution). After flushing the cells twice with PBS, they were fixated and permeabilized in fixaton/permeabilization buffer and stained with FoxP3 PE or isotype IgG1 kappa PE antibody (45 min in the dark at room temperature). Afterward, cells were washed twice with PBS, fixed in 300 μL of 1% paraformaldehyde in PBS and were examined by flow cytometry. Ten thousand counts of CD4 PerCP positive cells finished the cytometric acquisition. In the flow cytometry data analysis, two gate restrictions were used: R1 – FSC/SSC lymphocytes and R2 – CD4 PerCP positive cells (Fig. 1a, b). Next, common parts of R1 and R2 (R1*R2) were used for the examination of the results obtained. $CD4^{+}/CD25^{high}$ was defined as R3 (Fig. 1c), $CD4^{+}/CD25^{high}/CD127^{low}$ as R4 (Fig. 1d), and finally $CD4^{+}/CD25^{high}/CD127^{low}/FoxP3^{+}$ (nTreg) as R5 (Fig. 1e). Lymphocyte phenotype determination was conducted with FACS Calibur and BD CellQuestTM PRO software (BD Biosciences, Poland).

2.3 IgE Determination

Total IgE concentration was determined in serum samples by a solid phase enzyme-linked immunosorbent assay (ELISA) based on the sandwich principle, calibrated with commercially available IgE standards. Total IgE values are expressed in IU/ml.

2.4 Vitamin D Determination

An Dia-Sorin Liaison® analyzer (Saluggia, Italy) was used to determine the concentration of total 25-hydroxy vitamin D. The apparatus measures the concentration range from 4.0 to 150 ng/ml (in the SI unit: ng/ml × 2.5 = nmol/l). Values below 4.0 ng/ml are recorded as <4 ng/ml. The highest recorded value without dilution is 150 ng/ml. This test is a direct competitive assay technology using chemiluminescent immunoassays (CLIA) to quantify the level of 25-hydroxy vitamin D in serum and plasma. During the first incubation, 25-hydroxy vitamin D is separated from the binding protein and combined with an antibody to a solid phase. Vitamin D marker is added after 10 min. Subsequently, there is an incubation for 10 s, whereupon unbound material is removed in the rinse cycle. Then, initiating factors, which immediately cause a chemiluminescent reaction, are added sequentially. A light signal is measured using a photomultiplier, in relative light units (RLU). The signal is inversely proportional to the concentration of 25-hydroxy vitamin D present in the calibrators, control, and samples.

2.5 Statistical Elaboration

Age of patients was given as medians with lower and upper quartiles. Other variables, such as vitamin D content, and percentages of lymphocytes and white blood cells, were presented as means ± SD. Differences between groups were evaluated with a *t*-test (values with normal distribution) or Kruskal-Wallis test (values with non-parametric distribution). An analysis of the correlation between immunological indices and vitamin D content or disease severity have also been done. A p-value < 0.05 was adopted to define the level of statistical significance.

3 Results

Mild atopic dermatitis was diagnosed in 18% of patients; 41% of children had an average course of disease and 41% had severe disease.

3.1 Lymphocytes

The mean percentages of lymphocytes (atopic dermatitis group: 44.7 ± 16.2%, control group: 39.8 ± 11.3%) and eosinophils (atopic dermatitis group: 6.2 ± 4.5%, control group: 3.5 ± 2.4%) tended to be greater in atopic dermatitis, but the differences failed to reach statistical significance ($p = 0.36$ and $p = 0.06$, respectively). Likewise, there was insignificant difference in total IgE concentration between the two groups due to a very large data scatter, although the IgE content tended to by greater in atopic dermatitis (atopic dermatitis: 2127.6 ± 6736.6, control group: 70.3 ± 72.4, IU/ml; $p > 0.05$). However, a significantly lower percentage of regulatory T cells was observed in the serum of children with atopic dermatitis compared with the control group ($p < 0.00006$) (Fig. 2) There were no significant differences concerning the other immunological parameters: phenotype CD3, CD4, CD8, CD4/CD8, CD19, CD16/56, natural killer T (NKT) cells, and anti-CD3 HLA -DR3 between the atopic dermatitis and control groups (Table 1). The study did not reveal any association between disease severity and the percentage of immunological indices.

3.2 Vitamin D

The mean serum vitamin D level was similar in the atopic dermatitis and control groups; 24.3 ± 7.0 and 24.5 ± 9.1 ng/ml, respectively, which is close to the 20 ng/ml low side limit of what is considered adequate for healthy children. The study revealed an inverse association between vitamin D content and the percentage of CD8 cells ($p < 0.05$; $r = 0.62$) in atopic dermatitis. The ratio of CD4/CD8 was proportional to vitamin D content in atopic dermatitis ($p < 0.05$; $r = 0.66$). However, we failed to reveal any other associations between vitamin D content and the other immunological indices investigated in atopic dermatitis. Nor was there any association between vitamin D content and disease severity (Table 2).

4 Discussion

Numerous studies have focused on a significant role of regulatory T cells (Tregs) in the formation

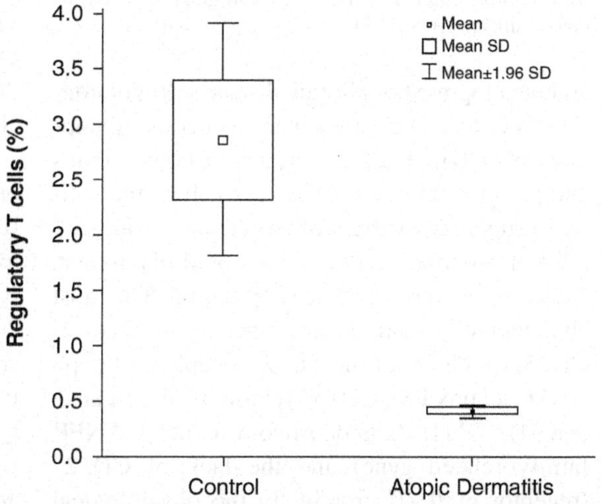

Fig. 2 Regulatory T cells in children with atopic dermatitis and healthy controls

Table 1 Content of indices investigated in atopic dermatitits and control groups

	AD group	Control group	p-value
IgE total (IU/ml)	2127.6 ± 6736.6	70.3 ± 72.4	ns
CD3 (%)	65.6 ± 9.7	61.3 ± 10.1	ns
CD19 (%)	19.7 ± 9.2	26.7 ± 12.4	ns
CD4 (%)	38.7 ± 7.6	34.4 ± 8.4	ns
CD8 (%)	25.4 ± 5.6	24.8 ± 10.6	ns
CD4/CD8 ratio	1.6 ± 0.6	1.7 ± 1.1	ns
CD16/56 (%)	9.6 ± 5.0	9.2 ± 3.9	ns
NKT (%)	2.1 ± 1.8	1.4 ± 1.4	ns
CD3 anty-HLADR3 (%)	6.7 ± 3.3	5.0 ± 4.5	ns
Regulatory T cells (%) ($CD4^+/CD25^{high}/CD127^{low}/FoxP3^+$)	0.41 ± 0.26	2.81 ± 2.19	**0.00001**
Lymphocytes (%)	44.7 ± 16.2	39.8 ± 11.3	ns
Monocytes (%)	7.4 ± 1.7	8.6 ± 2.2	ns
Granulocytes (%)	47.1 ± 16.2	50.9 ± 11.9	ns
Neutrophils (%)	40.7 ± 17.2	47.1 ± 12.2	ns
Eosinophils (%)	6.2 ± 4.5	3.5 ± 2.4	0.06

Data are means ± SD, *NKT* natural killer T cells, *ns* nonsignificant ($p > 0.05$)

Table 2 Associations between vitamin D content and immunological cellular indices in children with atopic dermatitis

	Vitamin D	p-value
CD3	−0.09	ns
CD19	0.39	ns
CD4	0.31	ns
CD8	−0.62	**$p < 0.05$**
CD4/CD8 ratio	0.66	**$p < 0.05$**
CD16/56	-0.42	ns
NKT	−0.39	ns
CD3 anty-HLA DR3	−0.43	ns
Tregs	0.36	ns

NKT natural killer T cells, *Tregs* T-regulatory cells, *ns* nonsignificant ($p > 0.05$)

and development of allergic diseases. Regulatory T cells $CD4^+CD25^+$ are a heterogeneous subpopulation of lymphocytes formed in both thymus and peripheral tissues. These cells show the ability to suppress a variety of target cells. There are a lot of proposed markers that could distinguish Tregs from the lymphocyte population. The most characteristic markers are the occurrence of CD25 (α chain of the IL-2 receptor), Foxp3 (forkhead box P3), CD152 (cytotoxic T cell antigen 4), GITR (glucocorticoid-induced TNFR family related gene) and the lack of CD127 (receptor of IL-7) protein. In the physiological condition, Tregs have the ability to maintain immunological tolerance by modulation of $CD4^+$ T cells, $CD8^+$ T cells, macrophages, NK cells, and the dendritic cells function (Sledź-Gawrońska 2010; Bacchetta et al. 2007). In atopic dermatitis, a decreased amount and attenuated function of Tregs have been observed (Lesiak et al. 2012; Samochocki et al. 2012; Hijnen et al. 2009; Szegedi et al. 2009). Tregs' dysfunction may be caused by mutations of the nuclear factor Foxp3 (Lyons et al. 2015). Additionally, staphylococcal colonization of the skin of patients with atopic dermatitis may modulate Tregs' function (Boguniewicz and Leung 2011). The results of the present study confirm the participation of regulatory T cells in allergic diseases as the number of Tregs was significantly lower in children with atopic dermatitis.

It is well established that activated CD4 cells, producing Th2 cytokines such as IL-4, IL-5, and IL-13, play a key role in allergy. These cytokines lead to overproduction of IgE and eosinophils (Gittler and Shemer 2012; Brandt and Sivaprasad 2011). In the acute phase of atopic dermatitis, skin biopsies are marked by perivenular infiltrates, predominantly consisting of lymphocytes and occasional monocytes. Lymphocytes are primarily $CD4^+$, $CD45RO^+$, and CLA^+, which suggests their previous contact with antigens (Sampson 2001). Hamid et al. (1994) have shown the increased expression of IL-4 and IL-13, the cytokines produced by the CD4 cells, in skin specimens collected during the acute phase of atopic dermatitis. In the present study, the amount of CD4 cells also was greater in children with atopic dermatitis, although the increase failed to reach statistical significance.

Studies indicate a deficiency of IFN-γ, a component of the atopic dermatitis pathogenesis (Gros et al. 2011). It is known that $CD8^+$ cells are a source of IFN-α (Machura et al. 2008). CD8 T lymphocytes and T helper cells have recently been divided into two groups: the cells secreting Th1-like cytokines, known as Tc1, and the cells secreting Th2 cytokines, known as Tc2 (Romagnani 1997). A reduced level of Tc1 cells, which have a high cytotoxic capacity, may be associated with enhanced propensity for skin infections in patients with atopic dermatitis (Lacour and Hauser 1993). In addition, Tc1 cell deficiency leads to the overproduction of IgE due to Tc1 capacity to regulate immunoglobulins. On the other hand, the number of Tc2 cells is not reduced in atopic dermatitis, as the production of Th2 cytokines, like IL-4 and IL-10, increases. Such quantitative and functional disorders of $CD8^+$ T cells are conducive for IgE antibody excess and skin infections in atopic dermatitis patients (Novak and Leung 2016; Kemeny et al. 1994). In the present study, no significant difference in the level of $CD8^+$ cells between the two groups was reveal, which may be due to the heterogeneity of this subpopulation of lymphocytes and the superiority of Tc2 over Tc1 cells in atopic dermatitis.

The influence of vitamin D on atopic dermatitis is related to the presence of a nuclear receptor for vitamin D in various immune-related cells. Bikle (2009) has shown that vitamin D might affect the expression of genes responsible for the proliferation of dendritic cells, macrophages, and other antigen presenting cells. In addition, vitamin D inhibits synthesis of IL-12 and IL-23, thereby reducing the number of Th1 cells (Baeke et al. 2010). It has also been shown that vitamin D increases the expression of antiphlogistic cytokines, IL-4, IL-5, and IL-10, and leads to the differentiation of CD4 regulatory T-cells, all of which plays a significant role in reducing allergic reactions. A number of studies have demonstrated the influence of vitamin D on the course of atopic dermatitis. The results, however, are often contradictory. Javanbakht et al. (2011) have demonstrated that vitamin D supplementation alleviates symptoms of atopic dermatitis compared with placebo, using the SCORAD index of disease severity. Likewise, a Japanese cohort study has shown a reduced risk of atopic dermatitis in early childhood in children of mothers being supplemented with vitamin D during pregnancy (Miyake et al. 2010). In contrast, Bäck et al. (2009) have shown that vitamin D supplementation in infancy is associated with increased risk of atopic dermatitis at 6 years of age.

In the present study, we did not observe a significant correlation between the serum vitamin D content and severity of atopic dermatitis. However, we noticed an appreciable enhancing effect of a higher content of vitamin D on the CD4/CD8 ratio. That effect may counter the notion of vitamin D protection against allergic diseases, which should be further explored in bigger population samples.

In conclusion, this study demonstrates that regulatory T cells were much reduced in children with atopic dermatitis compared with healthy population, which confirms these cells' role in the pathogenesis of allergy. On the other side, we failed to confirm the beneficial influence of vitamin D on the severity of atopic dermatitis.

Acknowledgements The authors thank Mr. Piotr Murawski, Head of ICT Department, of the Military Institute of Medicine for assistance in the implementation of statistical analysis. The authors also thank the study participants and their parents. Supported by grant 1/8865 (323) of the Military Institute of Medicine.

Conflicts of Interest The authors declare no competing interests in relation to this article.

References

Bacchetta R, Gambineri E, Roncarolo MG (2007) Role of regulatory T cells and FOXP3 in human diseases. J Allergy Clin Immunol 120:227–235

Bäck O, Blomquist HK, Hernell O, Stenberg B (2009) Does vitamin D intake during infancy promote the development of atopic allergy? Acta Derm Venereol 89:28–32

Baeke F, Takiishi T, Korf H, Gysemans C, Mathieu C (2010) Vitamin D: modulator of the immune system. Curr Opin Pharmacol 10:482–496

Bieber T (2010) Atopic dermatitis. Ann Dermatol 22:125–137

Bikle DD (2009) Vitamin D and immune function: understanding common pathways. Curr Osteoporos Rep 7:58–63

Boguniewicz M (2014) Middleton's allergy: principles and practice. In: Adkinson NF, Bochner BS, Burks AW et al (eds) Atopic dermatitis, 8th edn. Saunders, Philadelphia, pp 540–564

Boguniewicz M, Leung DY (2011) Atopic dermatitis: a disease of altered skin barrier and immune dysregulation. Immunol Rev 242:233–246

Brandt E, Sivaprasad U (2011) Th2 cytokines and atopic dermatitis. J Clin Cell Immunol 2:110. doi:10.4172/2155-9899.1000110

Darsow U, Wollenberg A, Simon D, Taïeb A, Werfel T, Oranje A, Gelmetti C, Svensson A, Deleuran M, Calza AM, Giusti F, Lübbe J, Seidenari S, Ring J, European Task Force on Atopic Dermatitis/EADV Eczema Task Force (2010) ETFAD/EADV eczema task force 2009 position paper on diagnosis and treatment of atopic dermatitis. J Eur Acad Dermatol Venereol 3:317–328

Gittler J, Shemer A (2012) Progressive activation of Th2/Th22 cytokines and selective epidermal proteins characterizes acute and chronic atopic dermatitis. J Allergy Clin Immunol 130:1344–1354

Gros E, Petzold S, Maintz L, Bieber T, Novak N (2011) Reduced IFN-γ receptor expression and attenuated IFN-γ response by dendritic cells in patients with atopic dermatitis. J Allergy Clin Immunol 128:1015–1021

Hamid Q, Boguniewicz M, Leung DYM (1994) Differential *in situ* cytokine gene expression in acute versus chronic atopic dermatitis. J Clin Invest 94:870–876

Hanifin JM, Rajka G (1980) Diagnostic features of atopic dermatitis. Acta Derm Venereol 92:44–47

Hijnen D, Haeck I, van Kraats AA, Nijhuis E, de Bruin-Weller M, Bruinzeel-Kommen C, Knol E (2009) Cyclosporin a reduces CD4 + CD25+ regulatory T-cell numbers in patients with atopic dermatitis. J Allergy Clin Immunol 124:856–858

Javanbakht MH, Keshavarz SA, Djalali M, Siassi F, Eshraghian MR, Firooz A, Seirafi H, Ehsani AH, Chamari M, Mirshafiey A (2011) Randomized controlled trial using vitamins E and D supplementation in atopic dermatitis. J Dermatol Treat 22:144–150

Kalicki B, Lewicki S, Stankiewicz W, Jung A, Rustecka A, Turkowska M, Rutkowski P, Bodera P, Zdanowski R (2013) Examination of correlation between vitamin D3 (25-OHD3) concentration and percentage of regulatory T lymphocytes (FoxP3) in children with allergy symptoms. Cent Eur J Immunol 38:70–75

Kapoor R, Menon C, Hoffstad O, Bilker W, Leclerc P, Margolis DJ (2008) The prevalence of atopic triad in children with physician-confirmed atopic dermatitis. J Am Acad Dermatol 58:68–73

Kemeny DM, Noble A, Holmes BJ, Diaz-Sanchez D (1994) Immune regulation: a new role for the CD8 + T cell. Immunol Today 15:107–112

Lacour M, Hauser C (1993) The role of microorganisms in atopic dermatitis. Clin Rev Allergy 11:491–503

Lesiak A, Smolewski P, Sysa-Jedrzejowska A, Narbut J (2012) The role of T-regulatory cells and toll-like receptors 2 and 4 in atopic dermatitis. Scand J Immunol 76:405–410

Leung DYM (2000) Atopic dermatitis: new insights and opportunities for therapeutic intervention. J Allergy Clin Immunol 105:860–876

Lyons J, Milner J, Stone K (2015) Atopic dermatitis in children: clinical features, pathophysiology and treatment. Immunol Allergy Clin N Am 35:161–183

Machura E, Mazur B, Golemiec E, Pindel M, Halkiewicz F (2008) *Staphylococcus aureus* skin colonization in atopic dermatitis children is associated with decreased IFN-gamma production by peripheral blood CD4+ and CD8+ T cells. Pediatr Allergy Immunol 19:37–45

Mesquita K, Costa I, Igreja A (2013) Atopic dermatitis and vitamin D: facts and controversies. An Bras Dermatol 88:945–953

Miyake Y, Sasaki S, Tanaka K, Hirota Y (2010) Dairy food, calcium and vitamin D intake in pregnancy, and wheeze and eczema in infants. Eur Respir J 35:1228–1234

Novak N, Leung DYM (2016) Pediatric allergy: principles and practice. In: DYM L, Szefler SJ, Bonilla FA, Akdis CA, Sampson H (eds) Role of barrier dysfunction and immune response in atopic dermatitis, 3rd edn. Elsevier, Toronto, pp 440–441

Ong PY, Leung DYM (2006) Immune dysregulation in atopic dermatitis. Curr Allergy Asthma Rep 5:384–389

Reitamo S, Luger TA, Steinhoff M (2008) Textbook of atopic dermatitis, 1st edn. Informa Ltd., London

Romagnani S (1997) The Th1/Th2 paradigm. Immunol Today 18:263–266

Romański B (1998) Allergic diseases at the turn of the century – epidemiology, prevention, treatment. Alerg Astma Immunol 1:12–16. (in Polish)

Samochocki Z, Alifier M, Bodera P, Jeziorkowska R, Rosiak E, Jurkiewicz B, Glińska O, Gliński W, Stankiewicz W (2012) T-regulatory cells in severe atopic dermatitis: alterations related to cytokines and other lymphocyte subpopulations. Arch Dermatol Res 304:795–801

Sampson HA (2001) Atopic dermatitis: immunological mechanisms in relation to phenotype. Pediatr Allergy Immunol 12(Suppl 14):62–68

Śledź-Gawrońska B (2010) The role of CD4 + CD25+ regulatory T cells in immunopathologies. J Lab Diagn 46:147–153. (in Polish)

Szegedi A, Baráth S, Nagy G, Szodoray P, Gál M, Sipka S, Bagdi E, Banham AH, Krenács L (2009) Regulatory T cells in atopic dermatitis: epidermal dendritic cell cluster may contribute to their local expansion. Br J Dermatol 160:984–993

Vassallo MF, Camargo CAJ (2010) Potential mechanisms for the hypothesized link between sunshine, vitamin D, and food allergy in children. J Allergy Clin Immunol 126:217–222

Williams H, Flohr C (2006) How epidemiology has challenged 3 prevailing concepts about atopic dermatitis. J Allergy Clin Immunol 1:209–213

Advs Exp. Medicine, Biology - Neuroscience and Respiration (2017) 32: 91–104
DOI 10.1007/5584_2017_17

Published online: 30 April 2017

Synergistic Activity for Natural and Synthetic Inhibitors of Angiogenesis Induced by Murine Sarcoma L-1 and Human Kidney Cancer Cells

Barbara J. Bałan, Andrzej K. Siwicki, Krzysztof Pastewka, Urszula Demkow, Piotr Skopiński, Ewa Skopińska-Różewska, Sławomir Lewicki, and Robert Zdanowski

Abstract

Tumor angiogenesis is an important link in the process of tumor growth and metastasis. A number of substances with an anti-angiogenic activity has been described, but their efficiency remains low. Many researchers believe that a better therapeutic effect could be achieved using a cocktail of several anti-angiogenic agents, having different points of action. A lot of synthetic and natural products of plant and animal origin have anti-tumor and anti-angiogenic properties. The aim of the present study was to evaluate the effect of some combinations of angiogenesis inhibitors on the growth and neovascularization of murine sarcoma L-1 , and on angiogenesis induced in the mouse skin by grafting of human renal cancer. The influence of theobromine, sulindac and its metabolite sulindac sulfone, chlorogenic acid, and shark liver oil on the afferent and efferent

B.J. Bałan (✉)
Department of Immunology, Biochemistry and Nutrition, Warsaw Medical University, 3 Oczki Street, 02-007 Warsaw, Poland
e-mail: bbalan@wum.edu.pl

A.K. Siwicki
Department of Microbiology and Clinical Immunology, University of Warmia and Mazury, 13 Oczapowskiego Street, 10-957 Olsztyn, Poland

K. Pastewka
Department of Urology, Dr. Tytus Chałubiński Hospital, 4 Lekarska Street, 26-600 Radom, Poland

U. Demkow
Department of Laboratory Diagnostics and Clinical Immunology of Developmental Age, Warsaw Medical University, Warsaw, Poland

P. Skopiński
Department of Histology and Embryology, Center for Biostructure Research, Warsaw Medical University, 5 Chałubińskiego Street, 02-004 Warsaw, Poland

E. Skopińska-Różewska
Department of Pathomorphology, Center for Biostructure Research, Warsaw Medical University, 5 Chałubińskiego Street, 02-004 Warsaw, Poland

S. Lewicki and R. Zdanowski
Department of Immunology, Biochemistry and Nutrition, Warsaw Medical University, Oczki 3, 02-007 Warsaw, Poland

angiogenesis pathways was tested. Individually, all of these substances suppressed tumor growth and angiogenesis. Synergy was found for a combination of theobromine, sulindac, and chlorogenic acid (L-1 sarcoma tumor growth), and for theobromine with sulindac sulfone or with shark liver oil, which were given to the mice grafted with human renal cancer cells (angiogenesis). No synergistic effects were shown after preincubation with tumor cells and inhibitors.

Keywords

Angiogenesis • Anti-angiogenic agents • Human renal cancer • Metastases • Murine sarcoma • Neovascularization • Tumor cells • Tumor growth

1 Introduction

Tumor angiogenesis is an important link in the process of tumor growth and metastasizing. Cells of a growing tumor require a continuous nutrient supply that is accomplished by an increased blood flow. Nonetheless, central part of a growing tumor becomes hypoxic, which results in release of hypoxia-inducible factors (HIFs) that encode for synthesis of angiogenic growth factors, endothelial cell activation in adjacent blood vessels, and degradation of the extracellular matrix, as schematically shown in Figs. 1 and 2 (Saaristo et al. 2000).

Anti-angiogenic therapy represents one of the more promising approaches to anticancer treatment. As demonstrated previously, L-1 sarcoma is a perfect experimental model that may be used to assess the influence of various substances of synthetic and natural origin upon the tumor growth and its cells activity (Skopińska-Różewska et al. 2007). A variety of compounds present in plant-contained foods and beverages have anti-cancer properties (Ebrahim and El Sayed 2016; Li et al. 2016; Sagar et al. 2006a, b; Kandaswami et al. 2005; Kamei et al. 1996). The cellular mechanisms of these properties consist of receptors for growth factors, synthesis of pro-angiogenic factors, and cell proliferation and apoptosis. Vegetal inhibitors of angiogenesis, such as catechins, gallocatechins, methyloxantines, or triterpenoid ursolic acid are common dietary components present in herbs, tea, or chocolate. Fruits, e.g., grapes and apples are rich in resveratrol. Caffeic and chlorogenic acids are widely distributed in the plant world or are constituents of an easily available raw material, e.g., convallamaroside is obtained from rhizomes of the *Convallaria majalis* (Lily of the Valley) (Nartowska et al. 2005).

Skopiński et al. (2004) have demonstrated that some of these substances suppress neovascularization induced in the mouse model by intradermal injection of a serum obtained from patients with diabetic retinopathy. The strongest anti-angiogenic effect has been observed with the use of a mixture of a few natural components containing a sulfone derivative of sulindac, a non-steroid anti-inflammatory drug. These substances also suppress the angiogenic activity of vascular endothelial growth factor (VEGF), fibroblast growth factor (bFGF), and interleukin-18 (IL-18) in experiments performed in the same experimental model of cutaneous angiogenesis (Skopiński et al. 2005).

One of the most widespread plant compound is chlorogenic acid. Previously, we have described an inhibitory effect of this polyphenolic acid in *ex vivo* angiogenic activity of human ovarian cancer cells and in their *in vitro* synthesis of bFGF and VEGF (Bałan et al. 1999). More recently, other studies have demonstrated that this phenolic acid is a strong metalloproteinase -9 inhibitor (Jin et al. 2005), has anti-angiogenic effects on choroidal neovascularization (Kim et al. 2010), decreases retinal vascular

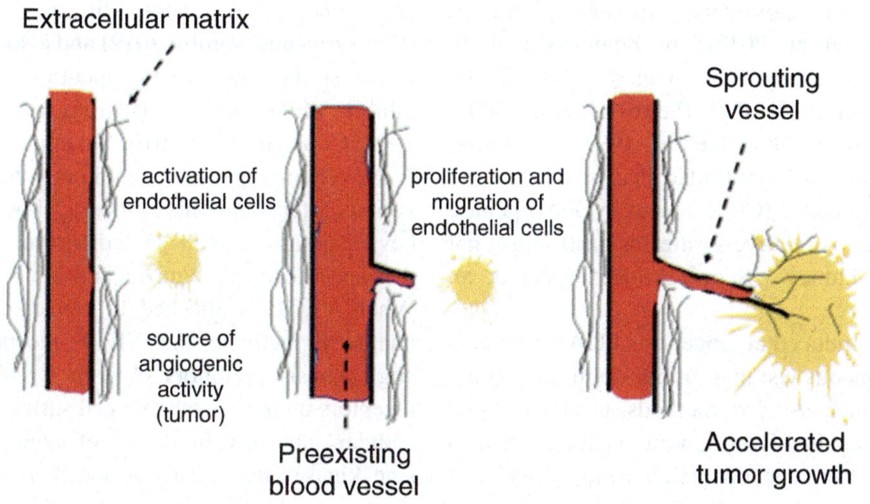

Fig. 1 Tumor angiogenesis stages. Newly-formed tumor-seeking blood vessel originated from the pre-existing one. In hypoxic tumor tissues, hypoxia-inducible factor-1α (HIF-1α) has a central role in inducing transcription of genes that are involved in angiogenesis, including vascular endothelial growth factor (VEGF)

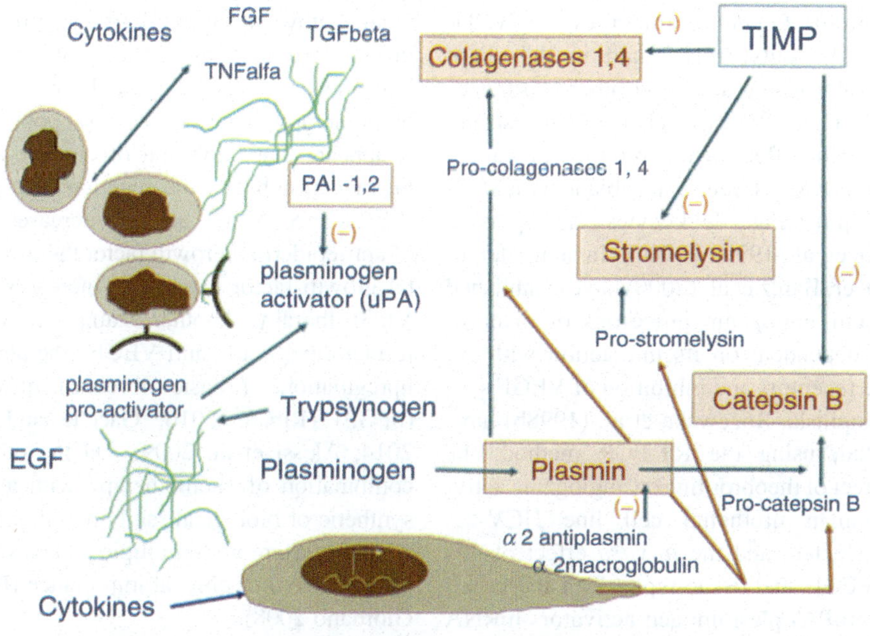

Fig. 2 Regulation of extracellular matrix degradation in early steps of angiogenesis. *EGF* vascular endothelial growth factor, *FGF* basic fibroblast growth factor, *TGFβ* transforming growth factor beta, *PAI* plasminogen activator inhibitor, *TNFα* tumor necrosis factor alpha, *TIMP* tissue inhibitors of metaloproteases, *PAI-1,2* plasminogen activator inhibitor-1 and 2, *uPA* urokinase-type plasminogen activator

hyperpermeability in the diabetic rat model (Shin et al. 2013), and suppresses angiogenesis *via* inhibition of HIF-1, leading to a reduction in VEGF expression (Park et al. 2015).

We have also reported that catechins, methyloxantines, alkyloglycerols, and some other natural compounds, and raw materials and remedies that contain them, have anti-angiogenic and growth-suppressing tumor properties (Skopiński et al. 2013a, b; Zdanowski et al. 2012; Skopińska-Różewska et al. 1999, 2003, 2011; Gibka et al. 2010; Pietrosiuk et al. 2004; Barcz et al. 1998; Gil et al. 1993). A similar effect is shown by sulindac that, not inhibiting cyclooxygenase 1 (COX-1) and cyclooxygenase 2 (COX-2), is relatively safe for gastrointestinal tract and kidneys (Skopińska-Różewska et al. 1998b).

Tumor-induced angiogenesis (TIA) is an efficient cutaneous test that, in our opinion, is superior to other *in vivo* methods used to assess angiogenesis involving laboratory mice, such as chorionic test, test on isolated skin flaps, test using implanted sponge or matrigel, and a corneal test. Using the TIA, we have reported the anti-angiogenic activity of sulindac and its metabolites, and of theobromine, catechins in cocoa tree seeds, salidroside and rosavin isolated from *Rhodiola* rhizomes, convallamaroside, alkyloglycerols found in the shark liver oil, chlorogenic acid, and other substances of natural origin. We have also shown a strong anti-angiogenic effect of theobromine on cancer cells (lung, ovary, kidney, and bladder) and on tumors infiltrating leukocytes (Skopińska-Różewska et al. 1999, 1998a). In a model of ovary cancer, Barcz et al. (2000) have established that the anti-angiogenic properties of theobromine are dependent on its interaction with A2 adenosine receptors and inhibition of VEGF synthesis. Skopińska-Różewska et al. (1998b) have investigated, using the RT-PCR method, the *in vivo* effect of theobromine on angiogenic activity of human urothelial cell line HCV-29, v-raf transfected, and the *in vitro* effect of this drug on VEGF, and tissue-type (tPA) and urokinase-type (uPA) plasminogen activators mRNA expression in this cell line. Theobromine suppresses angiogenesis, inhibites VEGF mRNA expression, but it has no effect on transcription of uPA and tPA. Bałan et al. (1999) have presented the inhibitory activity of caffeic and chlorogenic acids on angiogenic activity and VEGF and basic fibroblast growth factor (bFGF) synthesis in cultured ovary cancer cells. A key role of VEGF and its receptors in tumor angiogenesis is now firmly established (Rapisarda and Melillo 2012) and also confirmed in the studies on new therapeutics designed to inhibit VEGF activity (Petrelli and Giordano 2008; Cook and Figg 2010; Shi et al. 2016).

Novel angiogenesis inhibitors interfere with various steps in this process. For example, bevacizumab (Avastin®) is a monoclonal antibody that specifically recognizes and binds to VEGF. When VEGF is attached to bevacizumab, it is unable to activate the VEGF receptor. Other angiogenesis inhibitors bind to tyrosine kinase receptors on the endothelial cell surface, blocking cellular activities. Inhibition of angiogenic tyrosine kinases has been developed as a systemic treatment strategy for cancer. Three tyrosine kinase inhibitors, sunitinib, sorafenib, and pazopanib have been recently approved for treatment of patients with advanced cancer. Other small molecular inhibitors that target VEGF or other pathways of angiogenesis are currently investigated in clinical trials (Shi et al. 2016). Nonetheless, VEGF inhibitors, used in monotherapy, have yielded disappointing results. It appears that inhibition of one pro-angiogenic factor can enhance the function of another one, For instance, VEGF activity increases as a result of anti-epidermal growth factor therapy or placental growth factor expression increases after anti-VEGF therapy. Another example may be tumor refractoriness to anti-VEGF therapy due to upregulation of basic fibroblast growth factor (bFGF) (Ribatti 2016; Gacche and Meshram 2014; Alessi et al. 2009). On the other side, a combination of chemotherapy with at least two synthetic or biological inhibitors of angiogenesis, which interfere with multiple pathways of cancer growth, seem of clinical importance (Petrelli and Giordano 2008).

Yet novel multikinase inhibitors may exert a range of adverse effects related to the disturbance of VEGF-dependent physiological functions in the cardiovascular and renal systems, or having to do with wound healing and tissue repair. The appearance of side effects is particularly worrisome in case of renal cell carcinoma (RCC) as chemotherapy is of little relevance due to drug resistance of this most common renal cancer

(Beaumont et al. 2016; Boissier et al. 2016; Chen and Cleck 2009). RCC constitutes about 2–3% of all malignant tumors in adults. According to global data, RCC is diagnosed in nearly 270,000 persons and 116,000 patients die of it each year. In Europe, there are 65,000 new cases and over 25,000 deaths ascribed to RCC annually. The highest rate is observed in Eastern Europe and the lowest in Portugal and Spain. Men fall ill with RCC two times more often than women do. The incidence in Europe is 14.5/100,000 men and 6.9/100,000 women. In terms of morbidity, renal cancer is in the sixth place in men and ninth in women in Poland. RCC runs an aggressive course and has a poor prognosis due to local and systemic invasions. The most common histological type is clear cell renal cancer constituting 70–80% of all kidney cancers. For limited forms of RCC, nephrectomy is primary therapy. The treatment basis for advanced stages is angiogenesis inhibitors, although immunotherapy with interferon-α is also used (Rini et al. 2016).

Systemic anti-cancer treatment with inhibitors of angiogenesis is of rather limited value and is reserved primarily for patients with a high baseline severity of cancer, in which radical treatment is impossible or not possible, or for patients with recurrent disease (Choueiri et al. 2016; Tran et al. 2016). The aim of the present study was to evaluate the effects on sarcoma L-1 growth and angiogenesis induced in the mouse skin and on human renal cancer cells of theobromine, shark liver oil, chlorogenic acid, and sulindac and its metabolite sulfone in various combinations or individually.

2 Methods

2.1 Animals and Drugs

The study protocol was approved by a local Bioethics Committee and all institutional and national guidelines for the care and use of laboratory animals were followed. The study was performed in inbred Balb/c female mice, weighing 20–22 g, aged 8–10 weeks, delivered from the Polish Academy of Sciences breeding colony. The mice were housed 3–5 *per* cage and maintained under conventional conditions (room temperature of 22.5–23.0 °C, humidity of 50–70%, and a 12-h day/night cycle), with free access to standard rodent chaw and water *ad libitum*.

The following drugs were used: sulindac sulfone (OSI Pharmaceuticals; Melville, NY), sulindac, theobromine, chlorogenic acid (Sigma-Aldrich; Poznan, Poland), and shark liver oil (Ecomer®; ExposanAB, Aneby, Sweden).

2.2 Sarcoma L-1 Tumor Cells

Sarcoma L-1 cells from an *in vitro* culture stock were delivered from Warsaw's Oncology Center in Poland. The cells were passaged *in vivo* twice on syngeneic Balb/c mice and grafted into Balb/c mice subcutaneously (10^6/0.1 ml) into the subscapular region for the evaluation of tumor growth or intradermally for angiogenic activity. The mice were sacrificed with a lethal dose of pentobarbital (Morbital®; Biowet; Puławy, Poland) after 2 weeks, the tumors were excised, cut into small pieces, sieved, and suspended in 5 ml of PBS. The suspension was left for 10 min at room temperature. After sedimentation, the supernatant was collected and centrifuged for 10 min at 300 × *g*. The sarcoma cells obtained were washed once with PBS for 10 min, then centrifuged at 300 × *g* and resuspended in the Parker medium at a concentration of 10^7 cells/ml for tumor growth and 4 × 10^6 cells/ml for tumor-induced angiogenesis assays.

2.2.1 Subcutaneous Tumor Growth Assay

A suspension consisting of two million sarcoma cells was grafted subcutaneously into the mouse. On the day of cells grafting and on the following 13 days, the mice were fed with Eppendorf pipette-delivered theobromine 0.5 mg, or sulindac 0.2 mg, or chlorogenic acid 0.004 mg, or a cocktail of these drugs, or distilled water as a basal control. The mice were sacrificed after 2 weeks and the tumor mass was estimated.

2.2.2 Cutaneous Angiogenesis Assay for Tumor-Induced Angiogenesis (TIA) Test

Multiple 0.05 ml samples of 200,000 cells were injected intradermally into a partly shaved skin of Balb/c mice anesthetized with 3.6% chloral hydrate in a dose of 0.1 ml *per* 10 g of body mass (Sigma-Aldrich, Poznan, Poland), at least 3–4 mice *per* group. To facilitate the localization of an injection site later on, the suspension was colored with 0.1% of trypan blue. The mice were then fed with the above outlined drugs for 3 days, after which time they were sacrificed with a lethal dose of pentobarbital. All newly formed blood vessels on the inner skin surface were identified and counted in one-third of the microscopic central field (a dissection microscope at magnification of 6×). The identification was based on the fact that new blood vessels, directed to the site of cells injection, are thin and distinctly differ from the background vasculature in tortuosity and divarications.

2.3 Cutaneous Angiogenesis Induced in the Mouse by Cells Isolated from Human Kidney Cancer

Kidney cancer cells were obtained during nephrectomy from four patients with T2G2 claro-cellular renal cell carcinoma. Five gram of tumor tissue was dispersed mechanically, followed by enzymatic digestion with 0.1 mg/ml of collagenase (Sigma-Aldrich; Poznan, Poland) and 0.001 mg/ml of DNAse (Serva Electrophoresis GmbH, Heidelberg, Germany) dissolved in PBS, in a magnetic shaker at room temperature for 45 min. The cell suspension was then sieved, washed twice in PBS and suspended in the Parker medium at a concentration of 10^7/ml. Viability amounted to 97% of living cells as assessed with a 0.5% trypan blue exclusion test.

2.3.1 Experiments with Preincubation of Cells with Drugs

Cells, isolated as above described, in a concentration of 2×10^6/ml, were suspended in the Parker medium supplemented with L-glutamine, Hepes buffer, streptomycin (100 μg/ml) and penicillin (100 U/ml) in the presence of 5% fetal bovine serum (FBS) (Gibco™ media, Thermo Fisher Scientific, Waltham, MA) and with or without the addition of the following drugs: theobromine 10 and 20 μg per ml, sulindac sulfone 25, 50, and 100 μM, theobromine 10 μg/ml + sulindac sulfone 25 μM, and theobromine 10 μg/ml + sulindac sulfone 50 μM. The cultures were incubated for:

- 90 min at 37°C, equilibrated with a 5% CO_2/95% air mixture. After washing, samples of 5×10^5 cells in 0.05 ml were transplanted intradermally into the skin of Balb/c mice; at least four grafts in one mouse and at least 3 mice for one type of culture. Three days later the mice were euthanized and newly-formed blood vessels were counted on the inner skin surface, using a dissecting microscope;
- 24 h, washed, and suspended in a fresh culture medium without drugs and FBS. Then, incubated for another 24 h and the cells from supernatants were homogenized with an ultrasonic disrupter VirSonic 50 (Virtis; Gardiner, NY) for 2 min at 22.5 kHz, diluted four times with the Parker medium and multiple 0.05 ml samples were transplanted into the skin of Balb/c mice. Three days later, the mice were euthanized and newly-formed blood vessels were counted as above described.

2.3.2 Experiments without Preincubation

Tumor cells were isolated, homogenized, and transplanted into the skin of Balb/c mice. Then, mice received daily for the following 3 days:

- theobromine 100 μg in 0.2 ml of PBS subcutaneously, or sulindac sulfone 25 μg in 40 μl of distilled water by gavage, or both drugs;
- theobromine 100 μg in 0.2 ml of PBS, or Ecomer 0.01 ml, or both drugs.

At the end of third day, mice were euthanized and newly-formed blood vessels were counted on the inner skin surface as described above.

2.4 Statistical Analysis

Data were expressed as means ± SE. Statistical evaluation was performed using an unpaired *t*-tests and one-way ANOVA, followed by a *post-hoc* Tukey test. An α-level of 0.05 was taken as indicating statistically significant inter-group differences. A Combination Index (CI) was used as the standard quantitative measure of drug combination effect, with CI = 1 corresponding to additive effect and CI < 1 to synergism according to Chou (2006). A commercial GraphPad Prism ver. 5 packet was used to carry out the tests (GraphPad Software, La Jolla, CA).

3 Results

Figures 3 and 4 demonstrate the results of experiments performed in the sarcoma L-1 model. There was a highly significant reduction in tumor mass in the mice that received single inhibitors of angiogenesis such as theobromine, sulindac, and chlorogenic acid for 2 weeks, with a synergistic effect in the mice receiving a mixture thereof. Likewise, these drugs significantly reduced the number of newly-formed blood vessels in the cutaneous angiogenesis assay. However, synergy after a mixture of drugs was not observed here.

In the experiments with human renal cancer cells, pre-incubated with sulindac sulfone or theobromine, suppressive effect was obtained for all concentrations applied, except sulfone 25 μM, during the shorter lasting incubation (90 min). However, synergy was not observed in the mice which obtained a mixture of drugs, irrespective of incubation duration (Figs. 5 and 6) μg/ml

The synergistic effect was obtained after administration of a mixture of inhibitors for 3 days in post-transplant homogenates of renal cancer cells, and after combinations of theobromine with sulindac sulfone and theobromine with Ecomer (Figs. 7 and 8).

4 Discussion

Sarcoma L-1 tumor was described in the lung of a Balb/c mouse in 1976 and has since been maintained in serial passages *in vivo* and was also adapted to grow *in vitro* (Janik 1976). At present, L-1 sarcoma cells are stored in a cell

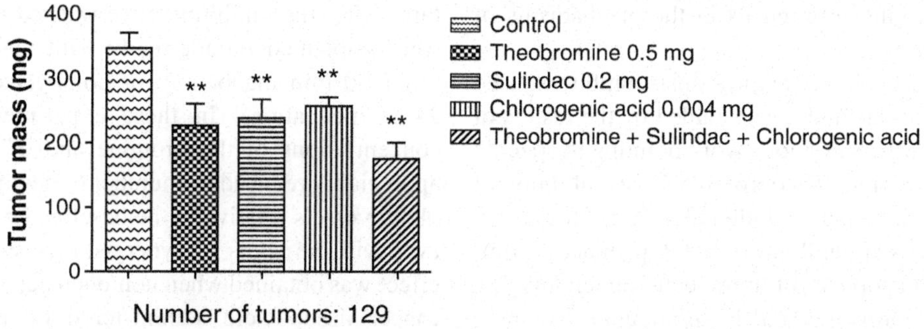

Fig. 3 **Synergistic effect on tumor mass** (CI = 0.73) of daily administration of theobromine, sulindac, and chlorogenic acid evaluated for 14 days after L-1 sarcoma cells transplantation into syngeneic Balb/c mice. Data are means ± SE; one-way ANOVA (p < 0.0001); **p < 0.01 for differences from the control value; *CI* combination index

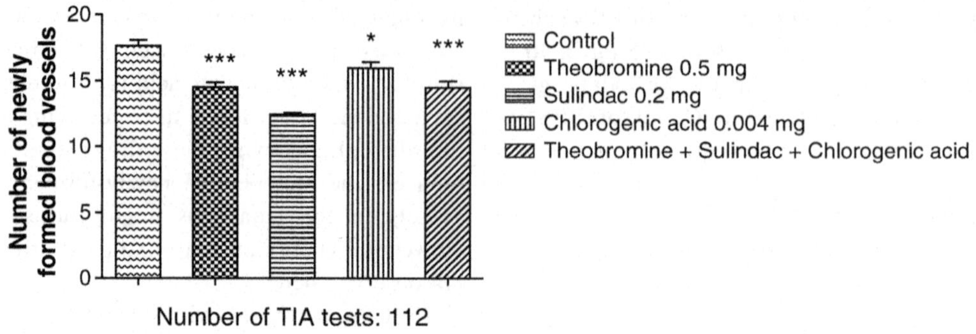

Fig. 4 **Lack of synergistic effect on the formation of new blood vessels** (CI = 1.02) of theobromine, sulindac, and chlorogenic acid, administered for 3 days to the mice with intradermal grafts of L-1 sarcoma cells. Data are means ± SE; one-way ANOVA ($p < 0.0001$); *$p < 0.05$; ***$p < 0.001$ for differences from the control value; *TIA* tumor-induced angiogenesis test, *CI* combination index

bank of the Oncology Center Cells Collection, in Warsaw, Poland. Using this tumor cell line, we have described anti-angiogenic effects of enoxaparin, *Aloe vera* extracts, various *Rhodiola* species, shark liver oil-based remedies, and a number of other herbal substances. In a study on PERVIVO, a multi-component digestive herbal remedy, we have reported a synergistic effect of this remedy and sulindac (Skopiński et al. 2013b). However, in the present study conducted in mouse L-1 sarcoma we failed to observe synergistic inhibitory effects on angiogenesis using a combination of several inhibitors such as theobromine, sulindac, and chlorogenic acid, despite a significant inhibition of angiogenesis exerted by each of them separately. On the other side, synergism was observed when tumor growth was evaluated, which is explicable by a longer period of administration of inhibitors and the emerging differences in the mechanisms of their action.

The selection of angiogenesis inhibitors, used in the present study, was guided by the results of our and others previous work pointing to specific sites of action. Theobromine is an inhibitor of VEGF path and endothelial cell proliferation, sulindac is an inhibitor of bFGF path and a stimulator of apoptosis of tumor cells and endothelial cells, chlorogenic acid is an antagonist of both VEGF and bFGF growth factors and a metallo-9-proteinase inhibitor (Sugimoto et al. 2014; Tonra and Hicklin 2007; Verhoef et al. 2006; Barcz et al. 2000; Barcz et al. 1998; Skopińska-Różewska et al. 1998a; D'Angelo et al. 1997; Gil et al. 1993; Moodie and Martin 1991; Leitman et al. 1986). Concerning the L-1 sarcoma mass, the strongest inhibitory effect was obtained for theobromine, and for angiogenesis inhibition it was for sulindac. The lack of synergism is possibly explicable by the overlapping contribution of individual inhibitors on production and release of various factors by tumor cells and the influence of these factors on target and effector cells. These phenomena should be considered in case of assays using living tumor cells. In the present study we attempted to sort out these phenomena in the experiments where we used the material obtained during nephrectomy performed in patients with renal cancer cells. In two experiments, tumor cells were incubated *in vitro* with individual inhibitors or their mixture. After the inhibitors were washed off, mice were implanted intradermally with viable cells (after 90 min incubation) or homogenate (after 24 h incubation). In these experiments, two concentrations of theobromine and of sulindac appreciably reduced the number of newly formed blood vessels relative to the baseline level; however with no signs of synergism. A synergistic effect was obtained when homogenates of kidney cancer tissue were administered to mice that were then treated with the inhibitors for the following 3 days. Synergism appeared in case of theobromine with sulindac and theobromine with

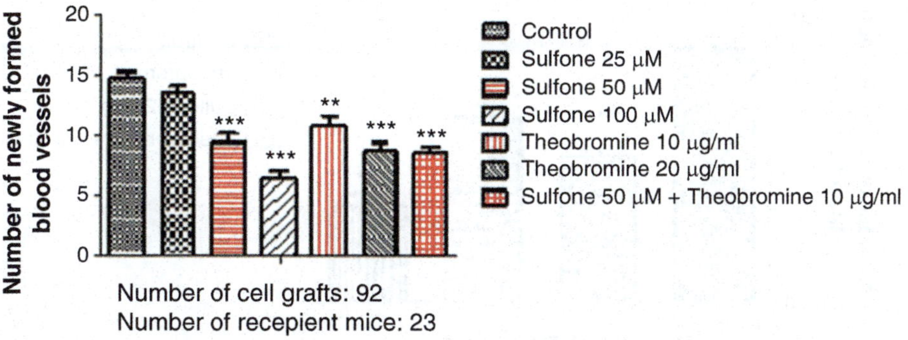

Fig. 5 **Lack of synergistic effect (CI = 0.84) of theobromine and sulfone on angiogenic activity** of human renal cancer cells. Cells were incubated with the drugs for 90 min, then washed and transplanted into the skin of Balb/c mice. Three days later, the mice were euthanized and newly-formed blood vessels were counted on the inner skin surface. Data are means ± SE; one-way ANOVA ($p < 0.0001$); **$p < 0.01$; ***$p < 0.001$ for differences from the control value; *CI* combination index

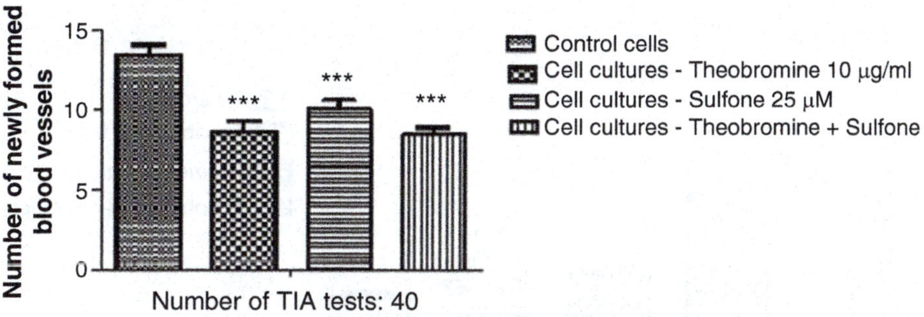

Fig. 6 **Lack of synergistic effect (CI = 0.91) of theobromine and sulfone on angiogenic activity** of human renal cancer cells. Cells were incubated with the drugs for 24 h, washed and incubated without drugs for the next 24 h. Then, cells with supernatants were homogenized, and transplanted into the skin of Balb/c mice. Three days later, the mice were euthanized and newly-formed blood vessels were counted on the inner skin surface. Data are means ± SE; one-way ANOVA ($p < 0.0001$); ***$p < 0.001$ for differences from the control value; *TIA* tumor-induced angiogenesis, *CI* combination index

Ecomer combinations. These results suggest that theobromine, beyond the known action on the production and release of cytokines by tumor cells, also acts on host environment and cellular targets in ways different from those of the other two anti-angiogenic agents, resulting in a hyper-additive synergism. The mechanism of theobromine action could have to do with the influence of c-AMP-dependent protein kinase A (cAMP/PKA) on the mitogenic action of VEGF and bFGF in endothelial cells, resulting in suppression of these cells proliferation by blocking Raf-1 kinase activation. The cytoplasmic Raf-1 kinase is essential for mitogenic signaling by growth factors that couple to tyrosine kinases, and by tumor-promoting compounds that activate protein kinase C (PKC) (D'Angelo et al. 1997). The mechanism of sulindac anti-angiogenic action, on the other side, could have to do with blocking the bFGF pathway and inducing endothelial apoptosis and cell cycle arrest *via* suppression of peroxisome proliferator-activated receptor (PPAR) transcriptional pathway (Li et al. 2013; Whitt et al. 2012; Jakubowska-Mućka et al. 2012; Liou et al. 2008; Flis et al. 2006; Skopiński et al. 2005; Elwich-Flis et al. 2003; Haanen 2001; Piazza et al. 2001, Rogala et al. 2000; Skopińska-Różewska et al.

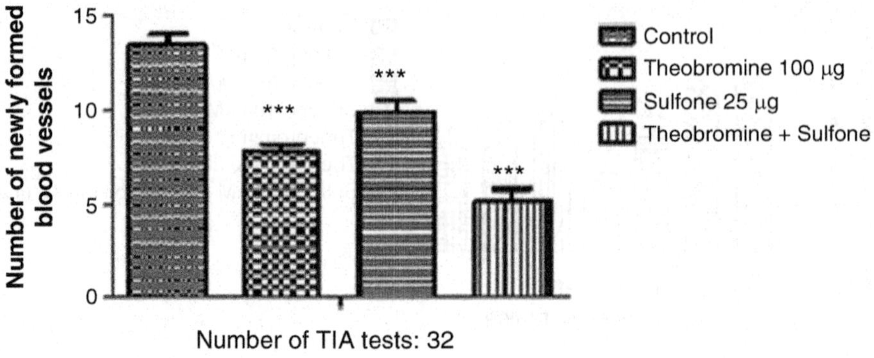

Fig. 7 Synergistic (CI = 0.58) suppressive effect of theobromine and sulfone on renal cancer cell homogenate-induced angiogenesis. Tumor cells were isolated, homogenized and transplanted into the skin of Balb/c mice. The mice were injected s.c. with theobromine, obtained sulfone by gavage, or both drugs, 100 and 25 μg, respectively, for the following 3 days. Then, mice were euthanized and newly-formed blood vessels were counted on the inner skin surface. Data are means ±SE; one-way ANOVA ($p < 0.0001$); ***$p < 0.001$ for differences from the control value (Tukey's *post hoc* test); *TIA* tumor-induced angiogenesis; *CI* combination index

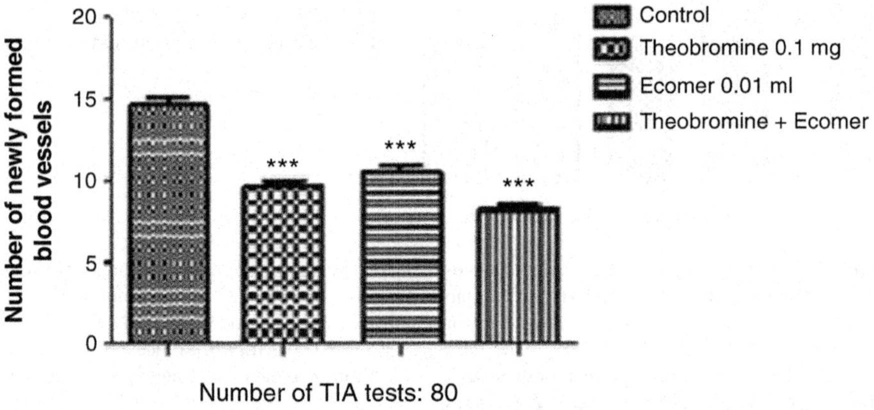

Fig. 8 Synergistic (CI = 0.78) suppressive effect of theobromine and Ecomer on renal cancer cell homogenate-induced angiogenesis. Tumor cells were isolated, homogenized, and transplanted into the skin of Balb/c mice. For the following 3 days mice were injected s.c. with 0.1 mg theobromine, obtained 0.01 ml Ecomer by gavage, or both drugs. Then, mice were euthanized and newly-formed blood vessels were counted on the inner skin surface. Data are means ± SE; one-way ANOVA ($p < 0.0001$); ***$p < 0.001$ for differences from the control value. Theobromine *vs.* Theobromine + Ecomer, $p = 0.0207$; Theobromine + Ecomer vs. Ecomer, $p = 0.0002$ (unpaired *t*-test); *TIA* tumor-induced angiogenesis; *CI* combination index

1998a). Sulindac increases the level of cellular c-GMP, activates c-GMP-dependent kinase and JNK kinase, and inhibits ERK 1/2, a cellular signal regulator. By modulating the activity of kinases, sulindac may induce apoptosis of cancer cells. Our previous study has shown that sulindac inhibits growth of transplanted tumors, made up of the L-1 sarcoma cell line, after a 14-day administration to Balb/c mice (Skopinska-Różewska et al. 2001). Another study has demonstrated an inhibitory effect of sulindac and its derivatives on tumor angiogenesis induced by kidney cancer cells (adenocarcinoma) and non-small cell lung carcinoma (Rogala et al. 2000).

In case of Ecomer, a preparation of shark liver oil rich in alkoxyglicerols (ether lipids), anti-angiogenic and anti-tumor effects could be connected to growth restrain and apoptotic activity (Pedrono et al. 2004; Skopinska-Różewska et al. 1999, 2003; Jackson et al. 1998; Vogler et al. 1998; Brohult et al. 1970). In the literature, there have been reports on synergistic effects of inhibitors of tumor growth (Pozdeyev et al. 2015; Wang et al. 2014; Yeong et al. 2014; Chougule et al. 2011; Li et al. 2009; Heider et al. 2008; Santini et al. 2006; Suganuma et al. 1999). Such studies, however, rarely tackle the issue of angiogenesis. Yang et al. (2003) have demonstrated that the antibiotic novobiocin has an anti-angiogenic activity that may be enhanced by a combination with vincristine. A synergistic effect of the heparanase inhibitor SST0001 in combination with anti-angiogenic agents (bevacizumab and sunitinib) has been observed in xenografts of pediatric sarcoma growth (Cassinelli et al. 2013). The present study adds up to those observations the synergistic anti-angiogenic effects of theobromine combined with sulindac sulfone and or with Ecomer.

Conflicts of Interest The authors declare no conflicts of interest in relation to this article.

References

Alessi P, Leali D, Camozzi M, Cantelmo A, Albini A, Presta M (2009) Anti-FGF2 approaches as a strategy to compensate resistance to anti-VEGF therapy: long-pentraxin 3 as a novel anti-angiogenic FGF2-antagonists. Eur Cytokine Netw 20:225–234

Bałan BJ, Skopińska-Różewska E, Barcz E, Sokolnicka I, Gawrychowski K, Strzelecka H (1999) The effect of selected phenolic acids in angiogenic activity of ovarian cancer cells. Onkol Pol 4:203–208. (Article in Polish)

Barcz E, Sommer E, Sokolnicka I, Gawrychowski K, Roszkowska-Purska K, Janik P, Skopińska-Różewska E (1998) The influence of theobromine on angiogenic activity and proangiogenic cytokines production of human ovarian cancers cells. Oncol Rep 5:517–520

Barcz E, Sommer E, Janik P, Marianowski L, Skopińska-Różewska E (2000) Adenosine receptor antagonism causes inhibition of angiogenic activity of human ovarian cancer cells. Oncol Rep 7:1285–1291

Beaumont JL, Salsman JM, Diaz J, Deen KC, McCann L, Powles T, Hackshaw MD, Motzer RJ, Cella D (2016) Quality-adjusted time without symptoms or toxicity analysis of pzopanib vesrsus sunitynib in patients with renal cell carcinoma. Cancer 122:1108–1115

Boissier E, Mir O, Hollebecque A, Izzedine H, Ederhy S, Gazzah A, Bahleda R, Missard C, Macquin-Mavier I, Tournigand C, Spano JP, Soria JC, Rousseau B (2016) Predictive factors of renal toxicities related to anti-VEGFR multikinase inhibitors in phase 1 trials. Invest New Drugs. Oct 25 (Epub ahead of print)

Brohult A, Brohult J, Brohult S (1970) Biochemical effects of alkoxyglicerols and their use in cancer therapy. Acta Chem Scand 24:730–735

Cassinelli G, Lanzi C, Tortoreto M, Cominetti D, Petrangolini G, Favini E, Zaffaroni N, Pisano C, Penco S, Vlodavsky I, Zunino F (2013) Antitumor efficacy of the heparanase inhibitor SST0001 alone and in combination with antiangigenic agents in the treatment of human pediatric sarcoma model. Biochem Pharmacol 15:1424–1432

Chen HX, Cleck JN (2009) Adverse effects of anticancer agents that target the VEGF pathway. Nat Rev Clin Oncol 6:465–477

Chou T-C (2006) Theoretical basis, experimental design, and computerized simulation of synergism and antagonism in drug combination studies. Pharmacol Rev 58:621–681

Choueiri TK, Escudier B, Powels T, Tannir NM, Mainwaring PN, Rini BI, Hammers HJ, Donskov F, Roth BJ, Peltola K, Lee JL, Heng DY, Schmidinger M, Agarval N, Sternberg CN, McDermott DF, Aftap DT, Hessel C, Scheffold C, Schwab G, Hutson TE, Pal S, Motzer RJ, METEOR investigators (2016) Cabozanitib versus everolimus in advanced renal cell carcinoma (METEOR): final results from the randomised, open-label, phase 3 trial. Lancet Oncol 17:919–927

Chougule M, Patel AR, Sachdeva P, Jackson T, Singh M (2011) Anticancer activity of noscapine, an opioid alkaloid in combination with cisplatine in human non-small cell lung cancer. Lung Cancer 73:271–282

Cook KM, Figg WD (2010) Angiogenesis inhibitors – current strategies and future prospects. CA Cancer J Clin 60:222–243

D'Angelo G, Lee H, Weiner RI (1997) cAMP-dependent protease inhibits the mitogenic action of vascular endothelial growth factor and fibroblast growth factor in capillary endothelial cells by blocking raf activation. J Cell Biochem 67:353–366

Ebrahim HY, El Sayed KA (2016) Discovery of novel anti-angiogenic marine natural product scaffolds. Mar Drugs 14(3):pii:E57; doi:10.3390/md14030057

Elwich-Flis S, Sołtysiak-Pawluczuk D, Spławinski J (2003) Anti-angiogenic and apoptotic effects of metabolites of sulindac on chick embryo chorioallantoic membrane. Hybrid Hybridomics 22:55–60

Flis S, Sołtysiak-Pawluczuk D, Jędrych A, Jastrzębski Z, Remiszewska M, Spławinski J (2006) Anti-angiogenic effect of sulindac sulfide could be secondary to induction of apoptosis and cell cycle arrest. Anticancer Res 26:3033–3042

Gacche RN, Meshram RJ (2014) Angiogenic factors as potential drug target: efficacy and limitations of anti-angiogenic therapy. Biochim Biophys Acta 1846:161–179

Gibka J, Wasiutyński A, Skopińska-Różewska E, Siwicki AK, Chorostowska Wynimko J, Sommer E, Mazurkiewicz M, Gliński M, Skurzak H, Wojcik R, Jung L (2010) The effect of undecanones and their derivatives on tumor angiogenesis and VEGF content. Pol J Vet Sci 13:105–115

Gil M, Skopińska-Różewska E, Radomska D, Demkow U, Skurzak H, Rochowska M, Beuth J, Roszkowski K (1993) Effect of purinergic receptor antagonists suramin and theobromine on tumor-induced angiogenesis in Balb/c mice. Folia Biol 39:63–68

Haanen C (2001) Sulindac and its derivatives: a novel class of anticancer agents. Curr Opin Investig Drugs 2:677–683

Heider U, von Metzler I, Kaiser M, Rosche M, Sterz J, Rotzler S, Rademacher J, Jacob C, Fleissner C, Kuckelkorn U, Kloetzel PM, Sezer O (2008) Synergistic interaction of the histone deacetylase inhibitor SAHA with the proteasome inhibitor bortezomib in mantle cell lymphoma. Eur J Hematol 80:133–142

Jackson JK, Burt HM, Octaba AM, Hunter W, Scheid MP, Mouchajir F, Lauener RW, Shen Y, Salari H, Duronio V (1998) The antineoplastic ether lipid, s-phosphonate selectively induces apoptosis in human leukemic cells and exhibits anti-angiogenic and apoptotic activity on the chorioallantoic membrane of the chick embryo. Cancer Chemother Pharmacol 41:326–332

Jakubowska-Mućka A, Sienko J, Świtaj T, Gołąb J, Lasek W (2012) Antitumor effect of sulindac in ovarian cell cultures. Ginekol Pol 82:195–199 (Article in Polish)

Janik P (1976) Lung colony assay in normal, irradiated and tumor bearing mice. Neoplasma 23:495–497

Jin UH, Lee JY, Kang SK, Park WH, Kim JG, Moon SK, Kim CH (2005) A phenolic compound, 5-caffeoylquinic acid (chlorogenic acid), is a new type and strong matrix metalloproteinase-9 inhibitor: isolation and identification from methanol extract of *Euonymus alatus*. Life Sci 77(22):2760–2769

Kamei H, Koide T, Kojimam T, Hasegawa M, Terabe K, Umeda T, Hashimoto Y (1996) Flavonoid-mediated tumor growth suppresion demonstrated by *in vivo* study. Cancer Biother Radiopharm 11:193–196

Kandaswami C, Lee LT, Lee PP, Hwang JJ, Ke FC, Huang YT, Lee MT (2005) The antitumor activities of flavonoids. In Vivo 19:895–899

Kim C, Yu HG, Sohn J (2010) The anti-angiogenic effect of chlorogenic acid on choroidal neovascularization. Korean J Ophthalmol 24:163–168

Leitman DC, Fiscus RR, Murad F (1986) Forskolin, phosphodiesterase inhibitors, and cyclic AMP analogs inhibit proliferation of culured bovine aortic endothelial cells. J Cell Physiol 127:237–243

Li H, Zhu X, Zhang Y, Xiang J, Chen H (2009) Arsenic trioxide exerts synergistic effect with cisplatin on non-small cell lung cancer cells via apoptosis induction. J Exp Clin Cancer Res 28:110. doi:10.1186/1756-9966-28-1101L1

Li N, Xi Y, Tinsley HN, Gurpinar E, Gary BD, Zhu B, Li Y, Chen X, Keeton AB, Abadi AH, Moyer MP, Grizzie WE, Chang WC, Clapper ML, Piaza GA (2013) Sulindac selectively inhibits colon tumor cell growth by activating the cGMP/PKG pathway to suppress Wnt/beta catenin signaling. Mol Cancer Ther 9:1848–1859

Li Q, Wang X, Dai T, Liu C, Li T, McClements DJ, Liu J (2016) Proanthocyanidins, isolated from *Choerospondias axillaris* fruit peels, exhibit potent antioxidant activities *in vitro* and a novel anti-angiogenic property *in vitro* and *in vivo*. J Agric Food Chem 64:3546–3556

Liou JY, Wu CC, Chen BR, Yen LB, Wu KK (2008) Nonsteroidal anti-inflammatory drugs induced endothelial apoptosis by perturbing peroxisome proliferator-activated receptor transcriptional pathway. Mol Pharmacol 74:1399–1406

Moodie SA, Martin W (1991) Effects of cyclic nucleotides and phorbol myristate acetate on proliferation of pig aortic endothelial cells. Br J Pharmacol 102:101–106

Nartowska J, Sommer A, Siwicki AK, Bany J, Skurzak H, Augustynowicz J, Skopińska- Różewska E (2005) The anti-angiogenic and immunotropic properties of convallamaroside, a steroidal saponin isolated from rhizomes and roots of *Convallaria majalis* L. Pol J Environ Stud 14:296–299

Park JJ, Hwang SJ, Park JH, Lee HJ (2015) Chlorogenic acid inhibits hypoxia-induced angiogenesis via down-regulation of the HIF-1 alpha/AKT pathway. Cell Oncol 38:111–118

Pedrono F, Martin B, Leduc C, Le Lan J, Saiag B, Legrand P, Moulinoux JP, Legrand AB (2004) Natural alkylglycerols restrain growth and metastasis of grafted tumors in mice. Nutr Cancer 48:64–69

Petrelli A, Giordano S (2008) From single- to multi-target drugs in cancer therapy: when aspecificity becomes an advantage. Curr Med Chem 15:422–432

Piazza GA, Thompson WJ, Pamukcu R, Alila HW, Whitehead CM, Liu L, Fetter JR, Gresh WE Jr, Klein-Szanto AJ, Farnell DR, Eto I, Grubbs CJ (2001) Exisulind, a novel proapoptic drug, inhibits rat urinary bladder tumorigenesis. Cancer Res 61:3961–3968

Pietrosiuk A, Furmanowa M, Skopińska-Rożewska E, Sommer E, Skurzak H, Bany J (2004) The effect of acetylshikonin isolated from *Lithospermum canescens* roots on tumor-induced cutaneous angiogenesis. Acta Pol Pharm 61:379–382

Pozdeyev N, Berlinberg A, Zhou Q, Wuensch K, Shibata H, Wood WM, Haugen BR (2015) Targeting the NF-KB pathway as a combination therapy for advanced thyroid cancer. PLoS One 10(8):e0134901. doi:10.1371/journal.pone.013

Rapisarda A, Melillo G (2012) Role of the VEGF/VEGFR axis in cancer biology and therapy. Adv Cancer Res 114:237–267

Ribatti D (2016) Tumor refractoriness to anti-VEGF therapy. Oncotarget 7(29):46668–46677

Rini BI, McDermott DF, Hammers H, Bro W, Bukowski RM, Faba B, Faba J, Figlin RA, Hutson T, Jonasch E, Joseph RW, Leibovich BC, Olencki T, Pantuck AJ, Quinn DI, Seery V, Voss MH, Vood CG, Atkins MB (2016) Society for Immunotherapy of Cancer consensus statement on immunotherapy for the treatment of renal cell carcinoma. J Immunother Cancer 4:81; e-Collection 2016

Rogala E, Skopińska-Różewska E, Sommer E, Sokolnicka I, Rudziński P, Caban P, Sawicka E, Pastewka K (2000) The effect of sulindac on angiogenic activity of human lung and kidney tumor cells. Onkol Pol 3:77–83 (Article in Polish)

Saaristo A, Karpanen T, Alitalo K (2000) Mechanism of angiogenesis and their use in the inhibition of tumor growth and metastasis. Oncogene 19:6122–6129

Sagar SM, Yance D, Wong RK (2006a) Natural health products that inhibit angiogenesis: a potential source for investigational new agents to treat cancer; part 1. Curr Oncol 13:14–26

Sagar SM, Yance D, Wong RK (2006b) Natural health products that inhibit angiogenesis: a potential source for investigational new agents to treat cancer; part 2. Curr Oncol 13:99–107

Santini D, Caragalia M, Vincenzi B, Holen I, Scarpa S, Budillon A, Tonini G (2006) Mechanism of disease: preclinical reports of antineoplastic synergistic action of bisphosphonates. Nat Clin Pract Oncol 3:325–338

Shi L, Zhou J, Wu J, Shen Y, Li X (2016) Anti-angiogenic therapy: strategies to develop potent VEGFR-2-tyrosine kinase inhibitors and future prospect. Curr Med Chem 23:1000–1040

Shin JY, Sohn J, Park KH (2013) Chlorogenic acid decreases retinal vascular hyperpermeability in diabetic rat model. J Korean Med Sci 28:608–613

Skopińska-Różewska E, Piazza GA, Sommer E, Pamukcu R, Barcz E, Filewska M, Kupis W, Caban R, Rudziński P, Bogdan J, Mlekodaj S, Sikorska E (1998a) Inhibition of angiogenesis by sulindac and its sulfone metabolite (FGN-1): a potential mechanism of their antineoplastic properties. Int J Tissue React 20:85–89

Skopińska-Różewska E, Janik P, Przybyszewska M, Sommer E, Białas-Chromiec B (1998b) Inhibitory effect of theobromine on induction of angiogenesis and VEGF mRNA expression in v-*raf* transfectants of human urethelial cells HCV-29. Int J Mol Med 2:649–652

Skopińska-Różewska E, Krotkiewski M, Sommer E, Rogala E, Filewska M, Białas-Chromiec B, Pastewka K, Skurzak H (1999) Inhibitory effect of shark liver oil on cutaneous angiogenesis induced in Balb/c mice by syngeneic sarcoma L-1, human urinary bladder and human kidney tumor cells. Oncol Rep 6:1341–1344

Skopinska-Różewska E, Białas-Chromiec B, Rogala E, Filewska M, Skurzak H (2001) Inhibitory effect of sulindac on L1 sarcoma growth in mice. Terapia 9 (3/2):25–27 (Article in Polish)

Skopińska-Różewska E, Chorostowska-Wynimko J, Krotkiewski M, Rogala E, Demkow U, Skurzak H (2003) Inhibitory effect of Greenland shark liver oil combined with squalen and arctic birch ashes on angiogenesis and L-1 sarcoma growth in Balb/c mice. Pol J Vet Sci 6:54–56

Skopińska-Różewska E, Skurzak H, Wasiutyński A, Jung L, Siwicki AK, Bałan BJ, Sommer E, Mazurkiewicz M, Skorupski M, Prątnicki A, Sokolnicka I, Skopiński P (2007) Sarcoma L-1 in mice as a model for the study of experimental angiogenesis. Cent Eur J Immunol 32:77–83

Skopińska-Różewska E, Wasiutyński A, Skopiński P, Siwicka D, Zdanowski R, Bodera P (2011) *In vivo* effect of two complex herbal remedies Echinasal and Bioaron C on antibody production and immunological angiogenesis in mice. Cent Eur J Immunol 3:139–144

Skopiński P, Szaflik J, Duda-Król B, Nartowska J, Sommer E, Chorostowska-Wynimko J, Demkow U, Skopińska-Różewska E (2004) Suppresion of angiogenic activity of sera from diabetic patients with non-proliferative retinopathy by compounds of herbal orgin and sulindac sulfone. Int J Mol Med 14:707–711

Skopiński P, Skopińska-Różewska E, Sommer E, Siwicki AK, Wasiutyński A (2005) The effect of some diet-derived angiogenesis inhibitors and sulindac sulfone on the ability of VEGF, bFGF and IL-18 to induce cutaneous neo-vascular response in mice. Pol J Environ Stud 14(Suppl II):325–329

Skopiński P, Zdanowski R, Bałan BJ, Siwicki AK, Kocik J, Lewicki S, Suska M, Pastewka K, Skopińska-Różewska E, Demkow U (2013a) *Aloe arborescens* And American cranberry (*Vaccinium macrocarpon*) extracts inhibit tumor-induced cutaneous angiogenesis in mice. Cent Eur J Immunol 38:480–485

Skopiński P, Bałan BJ, Kocik J, Zdanowski R, Lewicki S, Niemcewicz M, Gawrychowski K, Skopińska-Rozewska E, Stankiewicz W (2013b) Inhibitory effect of herbal remedy PERVIVO and anti-inflammatory drug sulindac on L-1 sarcoma tumor growth and tumor angiogenesis in Balb/c mice. Mediat Inflamm 2013:289789. doi:10.1155/2013/289789

Suganuma M, Okabe S, Kai Y, Sueoka N, Sueoka E, Fujiki H (1999) Synergistic effects of epigallocatechin gallate with epicatechin, sulindac, tamoxifen on cancer preventive activity in the human lung cancer cell line PC-9. Cancer Res 59:44–47

Sugimoto N, Miwa S, Hitomi Y, Nakamura H, Tsuchija H, Yachie A (2014) Theobromine, the primary methylxantine found in *Theobroma cacao*, prevents malignant glioblastoma proliferation by negatively regulating phosphodiesterase-4, extracellular signal-regulated kinase, act/mammalian target of rapamycin kinase, and nuclear factor-kappa B. Nutr Cancer 66:419–423

Tonra JR, Hicklin DJ (2007) Targeting the vascular endothelial growth factor pathway in the treatment of human malignancy. Immunol Investig 36:3–23

Tran TA, Leong HS, Pavia-Jimenez A, Fedyshyn S, Yang J, Kucejowa B, Sivanand S, Spence P, Xie XJ, Pena-Llopis S, Power N, Brugarolas J (2016) Fibroblast growth factor receptor- depended and -independend paracrine signaling by sunitinib-resistant renal cell carcinoma. Mol Cell Biol 36:1836–1855

Verhoef C, de Wilt JH, Verheul HM (2006) Angiogenesis inhibitors: perspectives for medical, surgical and radiation oncology. Curr Pharm Des 12:2623–2630

Vogler WR, Liu J, Volpert O, Ades EW, Bouck N (1998) The anticancer drug edelfosine is a potent inhibitor of neovascularisation *in vivo*. Cancer Investig 16:549–553

Wang LH, Li Y, Yang SN, Wang FY, Hou Y, Cui W, Chen K, Cao Q, Wang S, Zhang TY, Wang ZZ, Xiao W, Yang JY, Wu CF (2014) Gambogic acid synergistically potentiates cisplatin-induced apoptosis in non-small-cell lung cancer trough suppressing NF-KB and MAPK/HO-1 signaling. Br J Cancer 110:341–352

Whitt DJ, Li N, Tinsley HN, Chen X, Zhang W, Li Y, Gary BD, Keeton AB, Abadi AH, Grizzle WE, Piazza GA (2012) A novel sulindac derivative that potently suppresses colon tumor cell growth by inhibiting cGMP phosphodiesterase and beta catenin transcriptional activity. Cancer Prev Res 5:822–833

Yang J, Jiang M, Zhen YS (2003) Novobiocine inhibits and shows synergistic effect with wincristine. Xao Xue Bao 38:731–734 (Article in Chinese)

Yeong LT, Abdul Hamid R, Saiful Yazan L, Khaza'ai H, Awang Hamisin DE (2014) Synergistic action of compounds isolated from the hexane extract of *Ardisia crispa* root against tumor-promoting effect, *in vitro*. Nat Prod Res 28:2026–2030

Zdanowski R, Skopińska-Różewska E, Wasiutyński A, Skopiński P, Siwicki AK, Sobiczewska E, Lewicki S, Buchwald W, Kocik J, Stankiewicz W (2012) The effect of *Rhodiola kirilowii* extracts on tumor-induced angiogenesis in mice. Cent Eur J Immunol 37:131

Advs Exp. Medicine, Biology - Neuroscience and Respiration (2017) 32: 105–112
DOI 10.1007/5584_2017_2

Published online: 3 March 2017

Solitary Rectal Ulcer Syndrome in Children: A Case Series Study

K. Kowalska-Duplaga, I. Lazowska-Przeorek, K. Karolewska-Bochenek, M. Woynarowski, G. Czaja-Bulsa, A. Stawarski, S. Pieczarkowski, E. Hapyn, J. Jozefczuk, B. Korczowski, A. Szaflarska-Poplawska, and A. Banaszkiewicz

Abstract

Information on solitary rectal ulcer syndrome (SRUS) in children is limited and based on case reports only. This study was undertaken with the objective of describing the clinical history, symptoms, diagnostic work-up, and treatment of a large case series of pediatric patients with SRUS. The study was multi-center and retrospective. All pediatric endoscopists in Poland were invited to participate in the study and were asked to look through their endoscopic databases to identify SRUS cases from the last 10 years. The charts of SRUS patients were reviewed with respect to demographic data, and endoscopic and histological findings. Additionally, treatment methods and outcomes were assessed. In total, 31 patients (18 males, mean age of 13 years, range 5–18 years) were

K. Kowalska-Duplaga and S. Pieczarkowski
Department of Pediatric, Gastroenterology and Nutrition, Jagiellonian University Medical College, Cracow, Poland

I. Lazowska-Przeorek (✉), K. Karolewska-Bochenek, and A. Banaszkiewicz
Department of Pediatric Gastroenterology and Nutrition, Medical University of Warsaw, Warsaw, Poland
e-mail: ilazowska@gmail.com

M. Woynarowski
Department of Gastroenterology Hepatology and Feeding Disorders, Children's Memorial Health Institute, Warsaw, Poland

G. Czaja-Bulsa
Pediatric and Pediatric Nursery Unit of The Pomeranian Medical University in Szczecin, Szczecin, Poland

A. Stawarski
Department of Pediatrics, Gastroenterology and Nutrition, Medical University of Wroclaw, Wroclaw, Poland

E. Hapyn
Department of Pediatrics and Gastroenterology, Area Hospital in Torun, Torun, Poland

J. Jozefczuk
Pediatric Department with Children's Cardiological Subunit, Holy Spirit Hospital in Sandomierz, Sandomierz, Poland

B. Korczowski
Pediatric Department, Medical College, University of Rzeszów, Rzeszów, Poland

A. Szaflarska-Poplawska
Department of Pediatric Endoscopy and Gastrointestinal Function Testing, Collegium Medicum Nicolaus Copernicus University, Bydgoszcz, Poland

included in the study. All patients reported rectal bleeding. Other common symptoms included: abdominal pain (64.5%), perianal pain (54.8%), and passage of mucus (51.6%). The diagnostic work-up lasted from 1 to 48 months. Colonoscopic findings revealed rectal ulceration in 96.8% of patients. Therapeutic approaches included: high fiber diet (64.5%), laxatives (54.8%), topical corticosteroids (63.3%), 5-aminosalicylates administered orally and topically (29.1% and 96.8%, respectively), sucralfate (9.7%), and a biofeedback training (6.6%). Endoscopic argon plasma coagulation was performed in 2 patients and surgical intervention was necessary in 4 of them. Treatment was unsuccessful in 36% of patients. The findings of this study indicate that SRUS is rare in pediatric population, its diagnosis may be considerably delayed, and the treatment applied is often ineffective.

Keywords

Solitary rectal ulcer syndrome • Children • Rectal bleeding

1 Introduction

Solitary rectal ulcer syndrome (SRUS) is a rare disease in both adults and children. Although the pathophysiology of SRUS is unknown, there are several factors that could explain the presence of ulceration in the rectum. Firstly, SRUS could be a result of paradoxical contraction of pelvic floor (Mackle and Parks 1986). The excessive straining generates a high intrarectal pressure which pushes the anterior rectal mucosa into the contracting puborectalis muscle resulting in pressure necrosis of rectal mucosa. As a consequence, the anterior rectal mucosa is forced into the closed anal canal forming congestion, edema, and ulceration. The excessive straining may also result in local intussusceptions; another explanation of SRUS creation. Secondly, external anal sphincter may produce abnormal pressure gradients in the opposite direction which results in abnormal defecation leading to SRUS (Al-Brahim et al. 2009). Thirdly, there is growing evidence demonstrating thickened internal anal sphincter, abnormal balloon expulsion test, and abnormal sphincter relaxation as a typical feature in SRUS (Sharma et al. 2014; Halligan et al. 1995). Finally, direct trauma by repetitive self-digitation is also considered as a cause of SRUS. Interestingly, a significant group of patients with SRUS can benefit from behavioral therapy, which suggests the role of psychological factors in SRUS development. Obviously, coexistence of more than one of these pathogenic factors is liable to take place.

SRUS is children an area of limited knowledge that is mostly based on case reports. Therefore, the objective of this study was to describe the clinical history, symptoms, diagnostic work-up, and the treatment options applied in a large case series of pediatric patients.

2 Methods

This study was approved by the Clinical Research Ethics Committee of the Medical University of Warsaw, Poland. It was a retrospective, multi-center study. All 16 pediatric endoscopy centers in Poland were invited to participate. Each center received an email invitation and was asked to examine their endoscopic databases to identify SRUS cases in the 2005–2015 decade.

Table 1 Symptoms of SRUS at diagnosis

Symptom	n = 31(100%)
Rectal bleeding	31 (100%)
Abdominal pain	20 (64.5%)
Perianal pain	17 (54.8%)
Passage of mucus	16 (51.6%)
Constipation	14 (45.2%)
Diarrhea	14 (45.2%)
Straining at stool	14 (45.2%)
Sensation of incomplete defecation	11 (35.5%)
Rectal digitation	6 (19.4%)
Rectal prolapse	5 (16.1%)

SRUS Solitary rectal ulcer syndrome

The invitations were sent off three times at a 2-month interval. The patients' charts were reviewed with regard to age at diagnosis, sex, history of illness and clinical presentation, and the endoscopic and histological findings. Additionally, treatment methods and outcomes were assessed.

The identified patients consisted of children and teenagers under the age of 18 years, inpatients or outpatients diagnosed with SRUS and treated as such. Diagnosis of SRUS was based on characteristic endoscopic and histological findings. Lesions on the endoscopic findings were divided on the basis of number as solitary or multiple and on the basis of appearance as ulcerative, polypoidal/nodular, or erythematous mucosa only. The histological criteria included muscularization and fibrosis of the lamina propria with splayed smooth muscle extending between crypts, misshapen crypts, and thickened mucosal layer. Acute inflammation referred to the presence of neutrophils in the lamina propria and glands while chronic inflammation was characterized by lymphoplasmacytic infiltrate in the lamina propria.

3 Results

Nine out of the 16 endoscopy centers responded to the invitation to participate in the study. In total, 31 patients (F/M – 13/18, mean age 13 years, range 5–18 years) were identified. All patients reported rectal bleeding. Symptoms of SRUS are presented in Table 1. The whole diagnostic work-up lasted from 1 to 48 months (mean 10 ± 10(SD) months), including the time of symptoms duration before referral to the center where the final diagnosis was confirmed.

All children underwent colonoscopy with multiple biopsies (Table 2). Colonoscopic findings revealed various types of ulceration in 96.8% patients. Abnormal findings included: plain ulcers (32.2%), polyps with ulcers (22.6%), or a combination of different type of lesions. Singular ulcers were present in 19 (61.3%) patients but two of them had also polypoid lesions. Multiple ulcers (from 2 to 4) were described in six (25.8%) patients. In one case the only abnormality was erythema consisting of a flat, well-separated inflamed region. Other patients presented with a combination of different type of lesions. In all patients, endoscopic changes were located up to the sigmoid colon from the anal orifice, with the average distance of 8.7 cm. Biopsies showed different histological features consistent with diagnosis of SRUS.

In addition to colonoscopy and histological examination, 13% of patients underwent defecography and dynamic magnetic resonance (MRI). Anorectal manometry was done in 23% of patients. The results of these examinations supported diagnosis of SRUS.

Therapeutic approaches included the following: high fiber diet (64.5%), laxatives or stool softeners (54.8%), topical corticosteroids (63.3%), sulfasalazine or mesalazine administered orally and topically (29.1% and 96.8%, respectively), sucralfate (9.7%), and a biofeedback training (6.6%). Endoscopic argon plasma coagulation was performed in two patients and surgical intervention was necessary in four of them. Table 3 contains summary of applied therapeutic methods and their effectiveness. All therapeutic methods mentioned above were used separately or in different

Table 2 Endoscopic and histologic findings

No.	Gender	Symptoms duration (months)	Endoscopic findings: Location/ Distance from anal orifice (cm)	Endoscopic findings: Appearance	Histology
1	F	18	Rectum/NA	3 ulcers of up to 5 mm, polyp	Lamina propria fibrosis
2	M	6	Rectum/NA	4 ulcers of up to 5 mm	Lamina propria fibrosis, thickening of muscularis mucosa, focal superficial necrosis
3	F	2	Sigmoid colon, rectum/ 10 cm	1 ulcer of 30 mm	Vascular ectasia with inflammation
4	F	9	Rectum/NA	2 ulcers of 16 and 18 mm	Surface ulceration with inflammation
5	M	24	Rectum – posterior wall/NA	2 ulcers of up to 15 mm	Inflammatory infiltrate of lamina propria
6	M	6	Rectum/NA	1 ulcer of 20 mm	Surface ulceration with inflammation
7	M	16	Rectum/ 10 cm	Multiple polyps and erosions affecting ½ of rectum circumference	Lamia propria fibrosis, misshapen crypts, thickening of muscularis mucosa, inflammatory infiltrate of lamia propria, focal dysplasia
8	M	6	Rectum – anterior wall/ NA	Longitudinal ulcer of 25 mm; aggregation of several polyps	Misshapen crypts, inflammatory infiltrate of lamina propria
9	M	24	Rectum/4 cm	Several erosions affecting ½ of rectum circumference; polyp of 10 cm	Misshapen crypts, inflammatory infiltrate of lamina propria
10	F	3	Rectum – anterior wall/ NA	1 ulcer	Mixed inflammatory infiltrate
11	F	6	Rectum/8 cm	Flat, irregular ulceration	Inflammatory infiltrate of lamina propria
12	F	1	Rectum/1 cm	Deep ulceration of 8–10 mm	Inflammatory infiltrate of lamina propria
13	F	NA	Rectum/ 8–10 cm	Ulcer of 20 mm	Mild inflammation
14	F	1	Rectum/ 10–11 cm	Ulcer of 5–7 mm	Inflammatory infiltrate of lamina propria
15	M	24	Rectum/ 10 cm	Multiple polyps and erosions affecting in continuity of whole rectum circumference	Lamina propria fibrosis, misshapen crypts, thickening of muscularis mucosa, inflammatory infiltrate of lamina propria. Focal mild dysplasia in polyp
16	F	3	Rectum/NA	Ulcer of 10 mm	Inflammatory infiltrate of lamia propria
17	M	1.5	Rectum/NA	2 ulcers of 80 and 40 mm	Lamina propria fibrosis, thickening of muscularis mucosa, inflammatory infiltrate of lamia propria
18	M	12	Rectum/NA	Polyp with ulceration of 40 mm	Lamina propria fibrosis, inflammatory infiltrate of lamia propria
19	F	14	Rectum – anterior wall/ 8 cm	Ulcer of 14 × 16 mm	Lamina propria fibrosis, inflammatory infiltrate of lamina propria

(continued)

Table 2 (continued)

No.	Gender	Symptoms duration (months)	Endoscopic findings: Location/ Distance from anal orifice (cm)	Endoscopic findings: Appearance	Histology
20	M	5	Rectum – anterior wall/ 5 cm	3 ulcers of 10 × 8 mm, 6 × 8 mm, 6 × 6 mm	Lamina propria fibrosis, misshapen crypts, thickening of muscularis mucosa, inflammatory infiltrate of lamina propria
21	F	10	Rectum – anterior wall/ 8 cm	Ulcer of 10 × 8 mm	Lamina propria fibrosis, misshapen crypts, thickening of muscularis mucosa, inflammatory infiltrate of lamia propria
22	F	6	Rectum/NA	Ulcer of 30 × 30 mm	Lamina propria fibrosis, misshapen crypts. Granuloma
23	F	4	Sigmoid colon, rectum/NA	Ulcer of 40 × 30 mm	Lamina propria fibrosis, misshapen crypts; granuloma
24	F	4	Rectum/NA	2 polyps with ulcerations of 10 × 10 mm, 5 × 5 mm	Lamina propria fibrosis, inflammatory infiltrate of lamina propria
25	M	18	Rectum – posterior wall/NA	2 ulcers of 15 × 15 mm, 10 × 5 mm	Misshapen crypts, inflammatory infiltrate of lamina propria
26	M	48	Rectum – anterior wall/ 10 cm	1 flat lesion of 20 mm	Inflammatory infiltrate of lamina propria
27	M	4	Rectum/ 10 cm	1 ulcer of 10 mm	Inflammatory infiltrate of lamina propria
28	M	4	Rectum/8 cm	1 ulcer of 20 mm	Thickening of muscularis mucosa, inflammatory infiltrate of lamina propria
29	M	2	Rectum/8 cm	1 ulcer of 15 mm	Inflammatory infiltrate of lamina propria
30	M	11	Rectum – anterior wall/ 13 cm	1 ulcer of 30 × 25 mm	Inflammatory infiltrate of lamina propria
31	M	7	Rectum – anterior wall/ 9 cm	1 ulcer of 32 × 30 mm	Inflammatory infiltrate of lamina propria

NA data not available

constellations. Healing was confirmed endoscopically. Treatment was primarily unsuccessful or relapse occurred in 36% of patients.

4 Discussion

Recognition of SRUS in childhood is rare in Poland. There are no data on the prevalence of SRUS in this population. However, taking into consideration that between 2005 and 2015 approximately 32,000 colonoscopies were performed in children, even though we are not convinced we were able to report all SRUS patients due to a retrospective design of this study and possible misdiagnosis, we can estimate that SRUS is diagnosed in 1 per 1000 colonoscopies (Woynarowski et al. 2008, 2013). Other authors also underline a low frequency of this condition among both children and adults

Table 3 Treatments applied and their effectiveness in SRUS patients

Type of treatment	Patients treated (n)	Healing full or partial (n; %)
High fiber diet	20	9 (45.0)
Laxatives	17	9 (52.9)
Topical corticosteroids	19	12 (63.2)
Sulfasalazine or mesalazine topical	30	14 (46.7)
Sulfasalazine or mesalazine oral	9	7 (77.8)
Sucralfate	3	2 (66.7)
Biofeedback training	2	1 (50.0)
Endoscopic argon plasma coagulation	2	0 (0)
Surgical operation	4	4 (100.0)
Shortening of time in toilet	14	4 (42.8)
Avoiding of rectal digitation	5	2 (40.0)

SRUS solitary rectal ulcer syndrome

(Urgancı et al. 2013; Dehghani et al. 2012; Perito et al. 2012; Suresh et al. 2010; Al-Brahim et al. 2009; Ertem et al. 2002). Variability of symptoms, which obscures clinical picture and raises diagnostic difficulties, partially contributes to this situation. Clinical presentation often suggests other more common conditions as constipation or inflammatory bowel diseases. Differential diagnosis also includes infections, juvenile polyps, or sexual abuse. Family doctors or even gastroenterologists quite frequently do not suspect the diagnosis of SRUS, so that the condition may be misdiagnosed. In our group of patients, the main reason for hospitalization was rectal bleeding that seems the most common clinical manifestation of SRUS (Abid et al. 2012; Kc et al. 2008). Other symptoms such as abdominal pain, passage of mucus, constipation, diarrhea, straining at stool, feeling of incomplete defecation, and rectal digitation were recorded with similar frequency as in other reports (Blackburn et al. 2012). None of our patients were asymptomatic or presented with anal fissure. Asymptomatic individuals are mostly reported among adult population, as they are incidentally diagnosed during cancer-screening colonoscopy (Tjandra et al. 1992). Anal fissure may occur as a result of constipation and contribute to the blood loss (Al-Brahim et al. 2009), but no pediatric case report has described this complication. To summarize, we can confirm that in children, akin to adults, the most common symptom of SRUS is rectal bleeding.

Both clinical symptoms and endoscopic findings are not specific for SRUS, which underlines the necessity for histological diagnosis. Although all lesions in the series of cases herein presented were located in the rectum, some of them spread to the sigmoid colon. Moreover, not all of them were either solitary or ulcers as the condition name would suggest. Similar observation have been reported by other authors (Dehghani et al. 2012; Perito et al. 2012; Saadah et al. 2010; Ertem et al. 2002).

Histopathological findings, such as smooth muscle hyperplasia in lamina propria, hyperplasia of muscularis mucosae, surface ulceration, distortion of crypts architecture, or ectasia of superficial capillaries, were consistent with those previously described in pediatric SRUS cases (Dehghani et al. 2012; Perito et al. 2012). These changes are akin to those reported in the adult population. Some authors underline the difficulties in distinguishing SRUS from the inflammatory bowel disease (IBD) affecting only the distant rectum (Perito et al. 2012; Sharara et al. 2005). In SRUS, contrary to IBD, inflammatory infiltrate in lamina propria is not very advanced and muscular hyperplasia is common. Histological changes may facilitate final diagnosis, particularly in clinically atypical patients. Misdiagnosis may lead to the unjustified use of immunosuppressive drugs ineffective in SRUS treatment. On the other hand, some authors report a coexistence of SRUS and ulcerative colitis (Perito et al. 2012; Arhan et al.

2010). We failed to observe such a case in the group of patients presented. It is noteworthy that SRUS diagnosis, even in the absence of symptoms, always requires a full ileocolonoscopy with multiple biopsies.

Additional diagnostic procedures such as defecography, transrectal ultrasonography, dynamic MRI, or more recently endoscopic ultrasound, are reported useful in the diagnosis of SRUS. Both dynamic MRI and defecography enable to detect pelvic floor dysfunction, rectocele, and rectal intussusception or prolapse. Completeness of stool evacuation can be obtained only from the defecography. Anorectal manometry and electromyography provides the information about recto-anal inhibitory reflex, pressure profiles, defecation dynamics, rectal compliance, and the sensory threshold. These studies are so far rarely performed in the pediatric population. Our present retrospective study confirmed that observation as these procedures had been performed only in a handful of patients.

Blackbourn et al. (2012) have reported good results of behavior modification employed as adjunct treatment in SRUS. The improvement observed in 88% of patients consisted of fewer visits to the toilet, less time spend at defecation, and the encouragement not to persist with ineffective straining. Those authors have speculated that less tendency to repeatedly ineffectually strain breaks the vicious cycle of this condition. They also underline the need for maintaining compliance to prevent a relapse. All of our patients presented were advised to spend less time at defecation and to avoid straining and digital evacuation, but only few followed the doctor's instructions. The treatment, most frequently, consisted of topical 5-aminosalicylates or mesalamine. Another pharmacotherapy consisted of laxatives and stool softeners, oral 5-aminosalicylates or mesalamine, rectal corticosteroids, and sucralfate enemas. SRUS treatment is often unsuccessful, treatment modalities are not well established, and treatment data are meager in both children and adults. Until now, only have three prospective studies on SRUS been performed. None of those studies include children. Two observational studies have involved 12 patients treated with sucralfate enemas and 13 patients treated with a biofeedback (Malouf et al. 2001; Vaizey et al. 1997). Both sucralfate enema and biofeedback seem an effective initial treatment, but the long-lasting effect is questionable. In the only randomized trial to-date, 24 patients received a high-fiber diet, biofeedback therapy, and argon plasma coagulation (APC) or a high-fiber diet and biofeedback therapy alone (Somani et al. 2010). The authors have concluded that APC contributes to the healing process in SRUS patients.

The assessment of a specific treatment method is difficult as the treatments were used separately or in different combinations. Previous studies report a relief of symptoms achieved in some patients using a sucralfate enema (Dehghani et al. 2008; Zargar et al. 1991; Kochhar et al. 1990). In the present series of pediatric SRUS patients, enemas of sucralfate suspension were used only in three cases, with a good result. A biofeedback training, another previously employed therapeutic method in SRUS patients (Iwańczak et al. 2003; Vaizey et al. 1997), was used in the present study only in two patients, also with a good result. Endoscopic and surgical treatment were used occasionally; the latter provided a good effect. Although the present study describes the biggest group of pediatric SRUS cases yet, the number of patients remains still too small to create any guidelines. The main limitation of this study is a retrospective collection of data. Clinical aspects of SRUS in children, real effectiveness of treatment, and a way to improve the long-term follow-up care ought to be established in carefully planned prospective investigations; the more so that SRUS is a difficult-to-treat condition and relapses often occur.

In conclusion, this study demonstrates that the diagnosis of SRUS is rare in the pediatric population and may be considerably delayed. The need arises to increase the awareness of endoscopists and pathologists to consider the presence of this entity in differential diagnosis. There is a lack of controlled studies and standards of therapeutic approach to SRUS in children. Multi-center studies are required to explore details of clinical course and appropriate treatment in children suffering from SRUS.

References

Abid S, Khawaja A, Bhimani SA, Ahmad Z, Hamid S, Jafri W (2012) The clinical, endoscopic and histological spectrum of the solitary rectal ulcer syndrome: a single-center experience of 116 cases. BMC Gastroenterol 12:72

Al-Brahim N, Al-Awadhi N, Al-Enezi S, Alsurayei S, Ahmad M (2009) Solitary rectal ulcer syndrome: a clinicopathological study of 13 cases. Saudi J Gastroenterol 15(3):188–192

Arhan M, Onal IK, Ozin Y, Yesil Y, Temucin T, Ozdemir M, Ulker A (2010) Solitary rectal ulcer syndrome in association with ulcerative colitis: a case report. Inflamm Bowel Dis 16(2):190–191

Blackburn C, McDermott M, Bourke B (2012) Clinical presentation of and outcome for solitary rectal ulcer syndrome in children. J Pediatr Gastroenterol Nutr 54 (2):263–265

Dehghani SM, Haghighata M, Imanieha MH, Geramizadeh B (2008) Solitary rectal ulcer syndrome in children: a prospective study of cases from southern Iran. Eur J Gastroenterol Hepatol 20(2):93–95

Dehghani SM, Malekpour A, Haghighat M (2012) Solitary rectal ulcer syndrome in children: a literature review. World J Gastroenterol 18(45):6541–6545

Ertem D, Acar Y, Karaa EK, Pehlivanoglu E (2002) A rare and often unrecognized cause of hematochezia and tenesmus in childhood: solitary rectal ulcer syndrome. Pediatrics 110(6):e79

Halligan S, Sultan A, Rottenberg G, Bartram CI (1995) Endosonography of the anal sphincters in solitary rectal ulcer syndrome. Int J Color Dis 10(2):79–82

Iwańczak B, Blitek A, Krzesiek E (2003) Preliminary evaluation of the treatment of the children with solitary rectal ulcer, functional constipation and encopresis using biofeedback. Adv Clin Exp Med 12 (6):769–775

Kc S, Sharma S, Basnet B, Mishra AK (2008) Solitary rectal ulcer syndrome: uncommon cause of rectal bleeding in children. J Nepal Med Assoc 47 (172):238–240

Kochhar R, Mehta SK, Aggarwal R, Dhar A, Patel F (1990) Sucralfate enema in ulcerative rectosigmoid lesions. Dis Colon Rectum 33(1):49–51

Mackle EJ, Parks TG (1986) The pathogenesis and pathophysiology of rectal prolapse and solitary rectal ulcer syndrome. Clin Gastroenterol 15(4):985–1002

Malouf AJ, Vaizey CJ, Kamm MA (2001) Results of behavioral treatment (biofeedback) for solitary rectal ulcer syndrome. Dis Colon Rectum 44(1):72–76

Perito ER, Mileti E, Dalal DH, Cho SJ, Ferrell LD, McCracken M et al (2012) Solitary rectal ulcer syndrome in children and adolescents. J Pediatr Gastroenterol Nutr 54(2):266–270

Saadah OI, Al-Hubayshi MS, Ghanem AT (2010) Solitary rectal ulcer syndrome presenting as polypoid mass lesions in a young girl. World J Gastrointest Oncol 2 (8):332–334

Sharara AI, Azar C, Amr SS, Haddad M, Eloubeidi MA (2005) Solitary rectal ulcer syndrome: endoscopic spectrum and review of the literature. Gastrointest Endosc 62(5):755–762

Sharma A, Misra A, Ghoshal UC (2014) Fecal evacuation disorder among patients with solitary rectal ulcer syndrome: a case-control study. J Neurogastroenterol Motil 20(4):531–538

Somani SK, Ghosh A, Avasthi G, Goyal R, Gupta P (2010) Healing of solitary rectal ulcers with multiple sessions of argon plasma coagulation. Dig Endosc 22 (2):107–111

Suresh N, Ganesh R, Sathiyasekaran M (2010) Solitary rectal ulcer syndrome: a case series. Indian Pediatr 47 (12):1059–1061

Tjandra JJ, Fazio VW, Church JM, Lavery IC, Oakley JR, Milsom JW (1992) Clinical conundrum of solitary rectal ulcer. Dis Colon Rectum 35(3):227–234

Urgancı N, Kalyoncu D, Eken KG (2013) Solitary rectal ulcer syndrome in children: a report of six cases. Gut Liver 7(6):752–755

Vaizey CJ, Roy AJ, Kamm MA (1997) Prospective evaluation of the treatment of solitary rectal ulcer syndrome with biofeedback. Gut 41(6):817–820

Woynarowski M, Ignys I, Celinska-Cedro D et al (2008) Paediatric endoscopy in Poland in 2007. Pediatr Współcz Gastroenterol Hepatol Zywienie Dziecka 10 (3):158–163

Woynarowski M, Czkwianianc E, Respekta E et al (2013) Paediatric endoscopy in Poland in 2012. Pediatr Współcz Gastroenterol Hepatol Zywienie Dziecka 15 (2):47–54

Zargar SA, Khuroo MS, Mahajan R (1991) Sucralfate retention enemas in solitary rectal ulcer. Dis Colon Rectum 34(6):455–457

Advs Exp. Medicine, Biology - Neuroscience and Respiration (2017) 32: 113–122
DOI 10.1007/5584_2017_9

Published online: 28 February 2017

Knee Cartilage Regeneration with Umbilical Cord Mesenchymal Stem Cells Embedded in Collagen Scaffold Using Dry Arthroscopy Technique

B. Sadlik, G. Jaroslawski, D. Gladysz, M. Puszkarz, M. Markowska, K. Pawelec, D. Boruczkowski, and T. Oldak

Abstract

Articular cartilage injuries lead to progressive degeneration of the joint with subsequent progression to osteoarthritis, which currently becomes a serious health and economic issue. Due to limited capability for self-regeneration, cartilage repair remains a challenge for the present-day orthopedics. Currently, available therapeutic methods fail to provide satisfactory results. A search for other strategies that could regenerate a hyaline-like tissue with a durable effect and adequate mechanical properties is underway. Tissue engineering strategies comprise the use of an appropriately chosen scaffold in combination with seeding cells. Mesenchymal stem cells (MSC) provide an interesting new option in regenerative medicine with solid preclinical data and first promising clinical results. They act not only through direct cartilage formation, but also due to paracrine effects, such as releasing trophic factors, anti-inflammatory cytokines, and promoting angiogenesis. The MSC can be applied in an allogeneic setting without eliciting a host immune response. Out of the various available sources, MSC derived from Wharton's jelly of an umbilical cord seem to have many advantages over their counterparts. This article details a novel, single-staged, and minimally invasive technique for cartilage repair that involves dry arthroscopic implantation of scaffold-embedded allogenic mesenchymal stem cells isolated from umbilical cord Wharton's jelly.

B. Sadlik (✉), G. Jaroslawski, and M. Puszkarz
Biological Joint Reconstruction Department, St Luke's Hospital, Bielsko-Biala 43-300, Poland
e-mail: sadlik@lukasza.pl

D. Gladysz, M. Markowska, D. Boruczkowski, and T. Oldak
Polish Stem Cell Bank, Warsaw, Poland

K. Pawelec
Department of Pediatric Hematology and Oncology, Warsaw Medical University, Warsaw, Poland

Keywords
Articular cartilage • Cartilage reconstruction • Chondral defect • Matrix aided implantation • Mesenchymal stem cells • Tissue engineering • Wharton's jelly

1 Introduction

Cartilage degeneration is a significant and growing problem of modern orthopedics that creates a serious economic burden and becomes a growing public health issue. Even minor cartilage defects could lead to bone-on-bone contact and knee malalignment resulting in osteoarthritis, which is a leading cause of pain and disability among older population (Lawrence et al. 2008). With longer life expectancy, aging population, and increasing obesity, the prevalence of osteoarthritis appreciably rises (Cross et al. 2014). The goal of the successful regeneration is to preserve natural abilities of native cartilage including mechanical and functional properties of the knee as a weight-bearing joint. Moreover, the desired effect should be long-lasting. Achieving a successful reconstruction of cartilage tissue still remains a challenge due to its avascular and aneural nature that highly limits intrinsic regenerative capacity (Leijten et al. 2013). Several therapeutic approaches have been carried out to address cartilage regeneration, but they fail to fulfill clinical needs. Due to dissatisfying results, there is a need of employing novel, more effective therapeutic techniques. Recent advances in stem cell engineering have led to clinical application of various cell types with different methodological approaches and promising, yet conflicting results. Herein we present a novel method for cartilage regeneration with umbilical cord Wharton's jelly-derived mesenchymal stem cells, a collagen scaffold, and dry arthroscopy technique.

2 Scientific Rationale

2.1 Mesenchymal Stem Cells – Definition and Properties

Mesenchymal stem cells (MSC) are considered to be promising in tissue engineering. According to the guidelines the International Society for Cellular Therapy, MSC ought to be plastic-adherent during standardized culture conditions, express CD105, CD73, and CD90, be negative regarding lineage antigens (CD45, CD34, CD14, CD19, and HLA-DR), and be able to differentiate into osteoblasts, adipocytes, and chondroblasts (Dominici et al. 2006). They are considered weekly or non-immunogenic due to low expression of class I HLA and no expression of class II HLA and co-stimulatory molecules, and thus can be applied in an allogeneic setting (Law and Chaudhuri 2013). A study on MSC immunogenicity revealed that, when given intra-articularly to 5-year-old mares as an autologous, allogeneic, or even xenogeneic material, they elicit a host immune response only after re-exposure to xenogeneic cells. No arthroscopic or histologic changes in synovium have been detected (Pigott et al. 2013). Due to low immunogenicity, MSC in combination with biomaterials could constitute a tissue-engineered product available for off-the-shelf application.

2.2 Wharton's Jelly as Abundant Source of Mesenchymal Stem Cells

Wharton's jelly is a gelatinous substance composed out of high amount of extracellular matrix that surrounds and protects cord blood vessels (Wang et al. 2004). Although MSC can be isolated from various sources including bone marrow and adipose tissue, Wharton's jelly-derived MSC (WJ-MSC) seem to be a preferable source because of easiness and safety of the harvesting procedure as well as a rich number of cells contained in the umbilical cord. WJ-MSC have high proliferation and differentiation capabilities, superior to adult stem cell sources, and characteristics close to embryo-derived stem cells, but with no risk of potential tumorigenesis. In contrast, regenerative capabilities of adult stem cell sources seem to decrease with donor's age (Beane et al. 2014). There is evidence that WJ-MSC are genetically stable and retain their immature immunophenotype, functional features, and immunomodulatory properties during long-lasting ex-vivo expansion (Chen et al. 2014; La Rocca et al. 2013). Moreover, in comparison to other sources, MSC derived from neonatal tissues are immunologically privileged with high expression of immunomodulatory factors and low expression of class I HLA (Deuse et al. 2011).

2.3 Mechanism of Action and Preclinical Data on Cartilage Regeneration

MSC can influence cartilage regeneration trough differentiation into chondrogenic lineage, inducing proliferation and differentiation of chondrocyte progenitors, and modifying reaction of endogenous cells. Paracrine mechanisms also play an important role in enhanced regeneration, through release of trophic factors and exertion of anti-inflammatory effects (Toh et al. 2016). WJ-MSC down-regulate expression of matrix-degrading enzymes released from synovium and avert cartilage damage in the xenogeneic animal model (Saulnier et al. 2015). The WJ-MSC potential in treating cartilage lesions and the ability to differentiate into chondrogenic lineage was confirmed in a study with type 1 collagen hydrogel as a scaffold (Chen et al. 2013). These investigators have demonstrated the expression of cartilage-specific matrix proteins and a chondrogenic transcription factor after incubation in chondrogenic medium. The WJ-MSC capabilities to regenerate cartilage are comparable, or even superior, to other sources with the advantage of maintaining the immune-privileged characteristics (Danišovič et al. 2016; Liu et al. 2012; Wang et al. 2009). Both WJ-MSC and MSC derived from adult tissues are found compatible with various scaffolds, which leads to encouraging effects (Musumeci et al. 2014). Promising results regarding cartilage repair have already been reported in animal models and first clinical studies (Filardo et al. 2013).

3 WJ-MSC Procedures

3.1 Cell Culture Preparation

The therapeutic medical experiment consisting of tissue cultures of samples taken from human umbilical cord was approved by a local Bioethics Committee. Informed consent was obtained for sample collection from donor mothers scheduled for natural or cesarean delivery. Umbilical cord tissue was transported in a controlled temperature and processed within 48 h after delivery. Tissue fragments were washed in sterile saline with the addition of antibiotic/antimycotic solution, cut into 2 cm pieces (Fig. 1a), and had blood vessels removed (Fig. 1b). Subsequently, Wharton's jelly was sliced into 2 cm^3 scraps (Fig. 1c) that were placed in a flask for MSC growth in a xeno-free medium supplemented with antibiotics (Fig. 1d). Cell cultures were incubated in in the air with 5% CO_2 at 37 °C. After 2–3 weeks, tissue explants were removed and adherent cells were passaged until reaching 90% of confluence (Fig. 1e). Cells were reseeded at 1.2×10^4cells/cm^2 in culture flasks for further

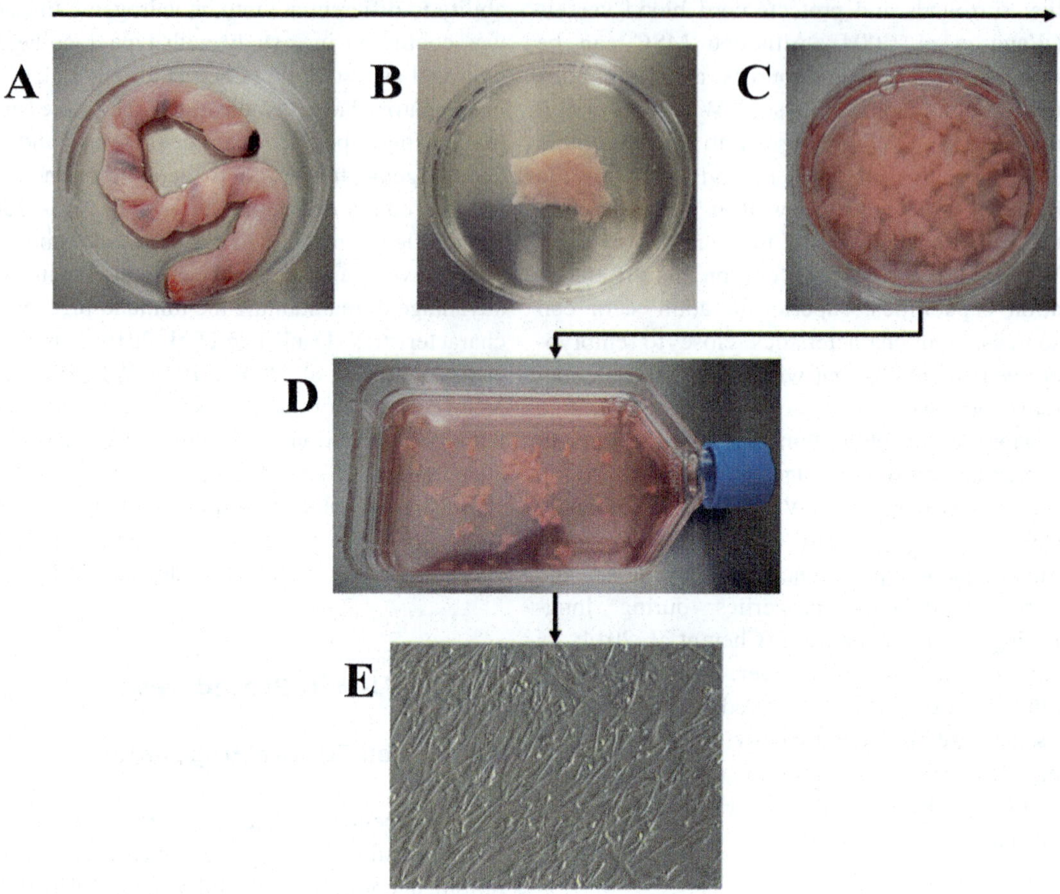

Fig. 1 **Key stages of cell culturing:** (**a**) – human umbilical cord (~10 cm); (**b**) – sectioned piece of Wharton's jelly after washing in sterile saline with antibiotic/antimycotic solution; (**c**) sectioned piece of Wharton's jelly sliced into 2 cm^3 fragments; (**d**) – scraps placed in a flask with xeno-free medium for mesenchymal stem cells (MSC) expansion; and (**e**) – a contrast image of Wharton's jelly MSC upon reaching 90% of confluence

expansion. Viability of expanded WJ-MSC was determined by the trypan blue exclusion in hemocytometer, and the cells' characteristics were confirmed with immunophenotyping by the presence or absence of surface markers (CD73-, CD90-, and CD105-positive; and CD34-, CD14-, CD19-, CD45, and HLA DR-negative). A reference sample of WJ-MSC was incubated with antibodies for 30 min in darkness and washed with cell wash solution. Next, cells were resuspended in cell fix solution and checked using flow cytometry with fluorescein isothiocyanate (FITC) and phycoerythrin (PE) conjugated to anti-mouse IgG1 antibody as control. The WJ-MSC intended for a therapeutic use were suspended in a mixture of human albumin in a presence of 10% dimethyl sulfoxide (DMSO), and transferred into freezing bags, which were placed in cell containers and cooled in a controlled rate freezer. After the freezing process, cells were stored in liquid nitrogen at −195 °C.

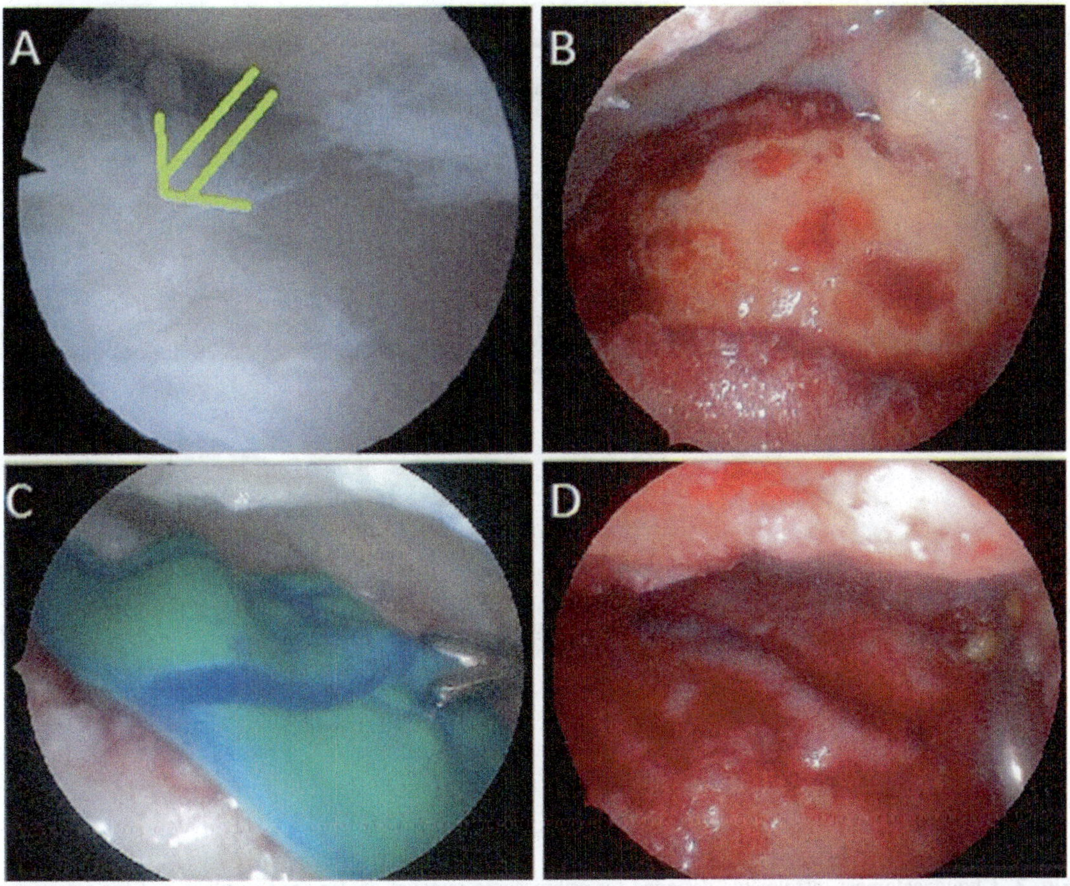

Fig. 2 The WJ-MSC procedure. Arthroscopic visualization of the right medial compartment through anterolateral working portal. (**a**) – full-thickness femoral condyle chondral injury (*arrow*); (**b**, **c**, and **d**) – retracting plate positioned to draw back capsule and adjacent synovial tissue to improve access to the chondral lesion: (**b**) – a view after preparing a chondral defect with vertical walls, (**c**) – a rubber templet positioned in the defect to check the shape and size of the scaffold, (**d**) – final position of a collagen scaffold embedded with WJ-MSC and covered with the fibrin glue to enhance stability of WJ-MSC in the graft

3.2 Thawing and Washing Procedure

The thawing procedure was conducted 30 min before the estimated implantation time. The freezing bag with WJ-MSC was quickly warmed up in a water bath at 37 °C. The DMSO cryoprotectant was washed out from WJ-MSC in a two-step dilution method with saline. The WJ-MSC were transferred to a sterile conical tube and suspended in 50 ml of saline. The cells were collected by centrifugation at 300 × *g* for 7 min at 22 °C. The supernatant was aspirated, and the pellet fraction was resuspended in 50 ml of saline and centrifuged again. The pellet fraction consisting of WJ-MSC was resuspended in 1 ml of saline. Finally, a suspension of WJ-MSC was transferred into a sterile syringe.

3.3 Patient Positioning and Arthroscopic Chondral Defect Preparation

The patient is positioned supine as for the standard knee arthroscopy. The procedure is typically performed under general or spinal anesthesia. A diagnostic arthroscopy is performed to visualize the entire cartilage injury and to characterize the lesions suitable for repair (Fig. 2a). Beside standard curettes, specially designed instruments

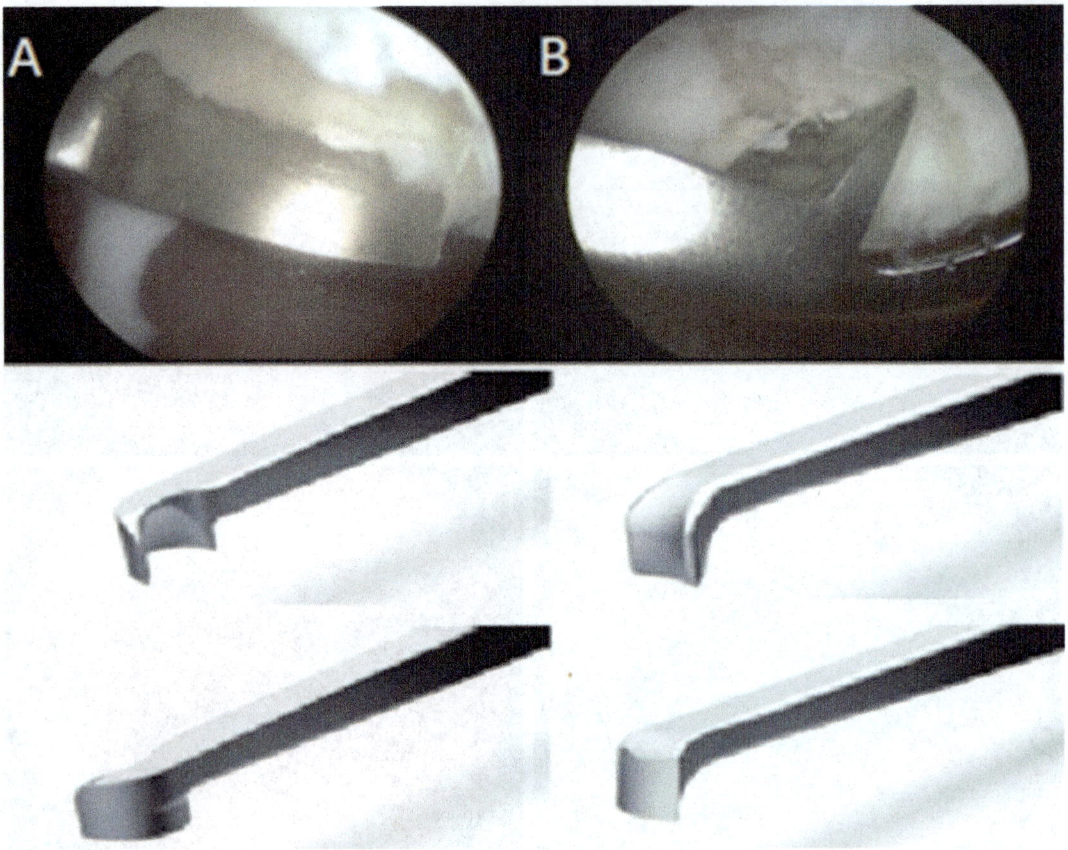

Fig. 3 **Chondrectomes:** instruments designed for arthroscopic removal of damaged cartilage and for creation of vertical ring surrounding the defected cartilage

(Chondrectomes Set, ATMED – Z. Rafalski, Katowice, Poland) may help optimize access and curettage of cartilage lesions, particularly when a parallel approach to the chondral defect is required (Fig. 3a, b). Loose chondral tissue associated with the lesion is excised, and curettes are used to create a contained lesion on the articular cartilage surrounded with a vertical border. Care is taken to remove the calcified cartilage layer overlying the subchondral bone, without violating the subchondral plate.

3.4 Wharton's Jelly-Derived Mesenchymal Stem Cell (WJ-MSC) Scaffold Preparation

A template created from aluminum foil or sterile latex dental dam (Sanctuary Dental Dam Systems; Ipoh, Malaysia) is inserted into the chondral defect to confirm the correct size and shape if needed (Fig. 2c). According to the defect dimensions, an appropriately sized implant is fashioned from the porcine type I/II collagen matrix (Chondrogide; Geistlich Biomaterials, Wolhusen, Switzerland) (Fig. 2d). The trimmed scaffold is wetted using saline and subsequently immersed with a suspension of WJ-MSC for 5 min, creating a malleable implant.

4 Dry Arthroscopic Implantation of WJ-MSC Embedded Scaffold

During dry arthroscopic visualization of the lesion, exposure is manageable by the retraction of the joint capsule and synovium, using a specially designed retracting system (Arthroscopic

Retracting System; ATMED – Z. Rafalski, Katowice, Poland) (Fig. 2b). When arthroscopic fluid is sucked from the working articular space, a skid or halfpipe has to be placed in the working portal to maintain an open gate to equalize the pressure in the joint. When undertaking an arthroscopic cartilage repair of a patella, femoral condyle, or tibial plateau, the entire cartilage defect should be visualized. The WJ-MSC embedded scaffold is inserted into the defect through the halfpipe and placed in the working portal with the use of a special inserter named 'fork'. Subsequently, the matrix is gently slid off from the 'fork' into the defect and kept in place, while an arthroscopic hook is introduced from the opposite portal to fit the implant into the prepared bed by pressing it (Fig. 2d). A fibrin glue (Tisseel Lyo; Baxter Healthcare Corp., Westlake Village, CA) is applied for covering the matrix to improve stability and to prevent WJ-MSC from migration into the synovial fluid. When the implant stability is confirmed, the wounds are closed and the joint is immobilized in the brace on the operating table.

4.1 Postoperative Rehabilitation and MRI Monitoring

The knee is immobilized for 5 days after the surgery to maintain a stable fibrin clot fully protecting biological implant. During this period patient is provided with muscle isokinetic training on the operated limb and exercise on the other body parts 4 times a day. A proper use of crutches is trained during first days after surgery. On the sixth day, the first passive mobilization with retracting the injured compartment is applied, followed by further passive mobilization 2–3 times a day. The MRI examination is carried out after the third week postoperatively to check for the graft position. When the examination confirms the proper status of the graft, more intensive mobilization is recommended. Weight-bearing is restricted for 3–6 weeks depending on the localization and size of the implanted scaffold. After 6 weeks, patient starts with a weight-bearing training of muscle strength, stability, and proprioception. Within the next 2–4 weeks, patient progresses to normal walk pattern without crutches. After 3 months, unrestricted physical activity is allowed. MRI scans are performed 1.5, 6, and 12 months after the surgery to observe graft incorporation and rebuilding. The early state of the chondral defect regeneration, with comparison to preoperative state, of the lateral compartment of a knee is presented in Fig. 4.

5 Discussion

Mesenchymal stem cells, in combination with biomaterials, carry a great potential that has already been proven in animal studies and first clinical applications. However, application of WJ-MSC have not yet been described in a clinical setting. With hitherto unsatisfactory results and a constantly increasing need for unicompartmental or total knee replacement surgery in case of large chondral lesions or osteoarthritic changes (Ackerman et al. 2016), a search for novel treatment methods has been underway. Tissue engineering strategies could constitute a major upturn in cartilage repair approaches. The key for successful articular cartilage regeneration lies in carefully selected components of tissue engineering: a scaffold with an adequate biomechanical properties compatible with non-immunogenic seeding cells with high chondrogenic potential, and a surgical technique adjusted for a precise graft implantation. Dry arthroscopy technique, along with scaffold and bone marrow-derived MSC, have been previously applied for the reconstruction of cartilage with good results (Gobbi et al. 2016, Gobbi et al. 2014). This technique requires a precise, minimally invasive surgery (Whyte et al. 2016; Sadlik and Wiewiorski 2014). The WJ-MSC procedure can be carried out regardless of the exact site of the knee cartilage injury due to the use of retracting plate, chondrectomes, and a halfpipe. The WJ-MSC is a preferable source of stem cells, with promising preclinical data. We propose a novel single-stage technique of cartilage regeneration that includes a suitably selected scaffold for WJ-MSC and dry arthroscopy technique, along with careful monitoring and controlled

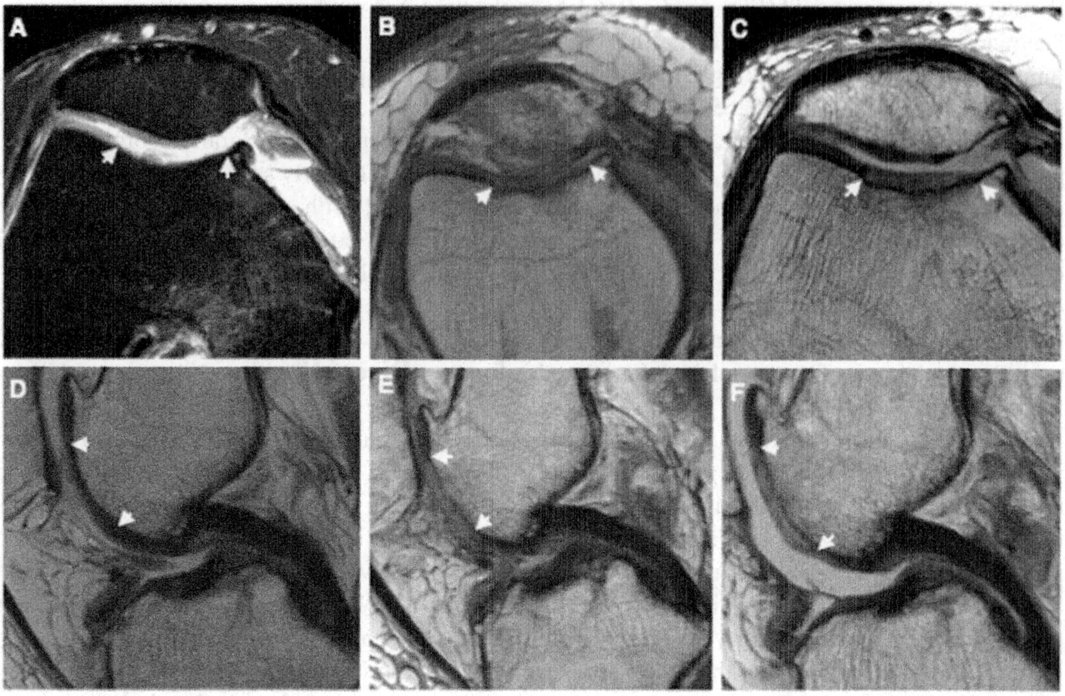

Fig. 4 Proton density MRI of a repaired defect (*arrows*): axial scans (*upper row*) and sagittal scans (*lower row*). (**a** and **d** – patellofemoral joint preoperatively: irregular cartilage defect of the trochlea (grade 3/4 according to the International Cartilage Repair Society (ICRS) scale of osteochondritis dissecan lesions), defect of medial femoral condyle, degenerative marginal spiking of the patellofemoral joint; (**b** and **e**) – regenerative tissue after the dry arthroscopic implantation of WJ-MSC embedded scaffold is well visible 3 weeks postoperatively (*green arrows*); (**c** and **f**) – after 6 weeks, regenerative tissue abounds and is integrated with the surrounding cartilage and subchondral lamina

rehabilitation. It is a challenging approach for cartilage regeneration, especially in patients with poor biological self-regenerative ability. MRI scans were done to assess the potential side effects and to evaluate the safety of proposed strategy. Synovial proliferation is considered a potential adverse effect of MSC-induced cartilage regeneration, as a large part of stem cells home in on the synovium rather than cartilage tissue. In view of such a risk, Koga et al. (2008) have suggested that MSC should be in a direct contact with cartilage surface for about 10 min. However, their study was based on the synovial MSC, which could have influenced the results. In the period from July 2015 to November 2016, we performed five surgeries, all approved by a local Bioethics Committee (permit no. 2015/06/25/1 BIL), in patients who had not benefited from standard therapies for a knee cartilage injury. We did not observe infections, excessive synovial proliferation, tumor formation, graft rejection, graft versus host reaction, or any other adverse effects. All patients also benefited from a significant knee pain reduction. However, long-term follow-up is required to assess the quality of newly formed cartilage and clinical outcomes.

This study has several limitations. A small number of patients in the preliminary phase of the study was a consequence of cautious patient qualification due to previous poor clinical experience with WJ-MSC application. After the promising results had been obtained with the very first cases and no apparent adverse events, the patient recruitment became more courageous. Another limitation was the lack of a control group, which was due to the fact that the study had been designed to confirm the efficiency of a new method on the basis of clinical results and MRI examinations. Moreover, it was a therapeutic experiment with the goal to achieve clinical benefits in patients who had not improved during standard therapy; thus a control group was unavailable. All enrolled patients met the criteria for total or unicompartmental knee alloplasty. The WJ-MSC provided the patients a chance to postpone the major surgical intervention and to substitute it for a minimally invasive treatment. The method described herein may be an important contribution to the advancement of tissue engineering aiming at articular cartilage repair, especially with regard to joint degenerative changes in elderly patients.

6 Conclusions

The aim of stem cell-enhanced cartilage repair is to acquire a hyalin cartilage-like tissue that is indistinguishable from the native one in both functional properties and histological structure. The WJ-MSC are promising seed cell candidates for tissue engineering and cell-based cartilage regeneration. Combined with an adequate scaffold, appropriate surgical technique, and a careful rehabilitation, they may hold the key for successful cartilage regeneration, particularly in elderly patients with poor intrinsic MSC regenerative potential. In this study we demonstrate that WJ-MSC could be used to induce regeneration of cartilage. To the best of our knowledge, this is the first presentation of clinical application of scaffold-embedded WJ-MSC through dry arthroscopy.

Supplementary Data Video presentation of arthroscopic cartilage repair of lateral compartment - https://www.youtube.com/watch?v=FPq_JU1DOskandfeature=youtu.be

Conflicts of Interest The authors declare no conflict of interest in relation to this article.

References

Ackerman IN, Bohensky MA, de Steiger R, Brand CA, Eskelinen A, Fenstad AM, Furnes O, Garellick G, Graves SE, Haapakoski J, Havelin LI, Mäkelä K, Mehnert F, Becic Pedersen A, Robertsson O (2016) Substantial rise in the lifetime risk of primary total knee replacement surgery for osteoarthritis from 2003 to 2013: an international, population-level analysis. Osteoarthritis Cartilage. http://dx.doi.org/10.1016/j.joca.2016.11.005

Beane OS, Fonseca VC, Cooper LL, Koren G, Darling EM (2014) Impact of aging on the regenerative properties of bone marrow-, muscle-, and adipose-derived mesenchymal stem/stromal cells. PLoS One 9(12):e115963. doi:10.1371/journal.pone.0115963

Chen X, Zhang F, He X, Xu Y, Yang Z, Chen L, Zhou S, Yang Y, Zhou Z, Sheng W, Zeng Y (2013) Chondrogenic differentiation of umbilical cord-derived mesenchymal stem cells in type I collagen-hydrogel for cartilage engineering. Injury 44(4):540–549

Chen G, Yue A, Ruan Z, Yin Y, Wang R, Ren Y, Zhu L (2014) Human umbilical cord-derived mesenchymal stem cells do not undergo malignant transformation during long-term culturing in serum-free medium. PLoS One 9(6):e98565. doi:10.1371/journal.pone.0098565

Cross M, Smith E, Hoy D, Nolte S, Ackerman I, Fransen M, Bridgett L, Williams S, Guillemin F, Hill CL, Laslett LL, Jones G, Cicuttini F, Osborne R, Vos T, Buchbinder R, Woolf A, March L (2014) The global burden of hip and knee osteoarthritis: estimates from the global burden of disease 2010 study. Ann Rheum Dis 73(7):1323–1330

Danišovič Ľ, Boháč M, Zamborský R, Oravcová L, Provazníková Z, Csöbönyeiová M, Varga I (2016) Comparative analysis of mesenchymal stromal cells from different tissue sources in respect to articular cartilage tissue engineering. Gen Physiol Biophys 35 (2):207–214

Deuse T, Stubbendorff M, Tang-Quan K, Phillips N, Kay MA, Eiermann T, Phan TT, Volk HD, Reichenspurner H, Robbins RC, Schrepfer S (2011) Immunogenicity and immunomodulatory properties of umbilical cord lining mesenchymal stem cells. Cell Transplant 20(5):655–667

Dominici M, Le Blanc K, Mueller I, Slaper-Cortenbach I, Marini F, Krause D, Deans R, Keating A, Dj P, Horwitz E (2006) Minimal criteria for defining multipotent mesenchymal stromal cells. The International Society for Cellular Therapy position statement. Cytotherapy 8(4):315–317

Filardo G, Madry H, Jelic M, Roffi A, Cucchiarini M, Kon E (2013) Mesenchymal stem cells for the treatment of cartilage lesions: from preclinical findings to clinical application in orthopaedics. Knee Surg Sports Traumatol Arthrosc 21(8):1717–1729

Gobbi A, Karnatzikos G, Sankineani SR (2014) One-step surgery with multipotent stem cells for the treatment of large full-thickness chondral defects of the knee. Am J Sports Med 42(3):648–657

Gobbi A, Scotti C, Karnatzikos G, Mudhigere A, Castro M, Peretti GM (2016) One-step surgery with multipotent stem cells and hyaluronan-based scaffold for the treatment of full-thickness chondral defects of the knee in patients older than 45 years. Knee Surg Sports Traumatol Arthrosc. doi:10.1007/s00167-016-3984-6

Koga H, Shimaya M, Muneta T, Nimura A, Morito T, Hayashi M, Suzuki S, Ju YJ, Mochizuki T, Sekiya I (2008) Local adherent technique for transplanting mesenchymal stem cells as a potential treatment of cartilage defect. Arthritis Res Ther 10(4):R84

La Rocca G, Lo Iacono M, Corsello T, Corrao S, Farina F, Anzalone R (2013) Human Wharton's jelly mesenchymal stem cells maintain the expression of key immunomodulatory molecules when subjected to osteogenic, adipogenic and chondrogenic differentiation in vitro: new perspectives for cellular therapy. Curr Stem Cell Res Ther 8(1):100–113

Law S, Chaudhuri S (2013) Mesenchymal stem cell and regenerative medicine: regeneration versus immunomodulatory challenges. Am J Stem Cells 2(1):22–38

Lawrence RC, Felson DT, Helmick CG, Arnold LM, Choi H, Deyo RA, Gabriel S, Hirsch R, Hochberg MC, Hunder GG, Jordan JM, Katz JN, Kremers HM, Wolfe F, National Arthritis Data Workgroup (2008) Estimates of the prevalence of arthritis and other rheumatic conditions in the United States, Part II. Arthritis Rheum 58(1):26–35

Leijten JC, Georgi N, Wu L, van Blitterswijk CA, Karperien M (2013) Cell sources for articular cartilage repair strategies: shifting from monocultures to cocultures. Tissue Eng B Rev 19(1):31–40

Liu S, Yuan M, Hou K, Zhang L, Zheng X, Zhao B, Sui X, Xu W, Lu S, Guo Q (2012) Immune characterization of mesenchymal stem cells in human umbilical cord Wharton's jelly and derived cartilage cells. Cell Immunol 278(1–2):35–44

Musumeci G, Castrogiovanni P, Leonardi R, Trovato FM, Szychlinska MA, Di Giunta A, Loreto C, Castorina S (2014) New perspectives for articular cartilage repair treatment through tissue engineering: a contemporary review. World J Orthop 5(2):80–88

Pigott JH, Ishihara A, Wellman ML, Russell DS, Bertone AL (2013) Investigation of the immune response to autologous, allogeneic, and xenogeneic mesenchymal stem cells after intra-articular injection in horses. Vet Immunol Immunopathol 156(1–2):99–106

Sadlik B, Wiewiorski M (2014) Dry arthroscopy with a retraction system for matrix-aided cartilage repair of patellar lesions. Arthrosc Tech 3(1):e141–e144. doi:10.1016/j.eats.2013.09.010

Saulnier N, Viguier E, Perrier-Groult E, Chenu C, Pillet E, Roger T, Maddens S, Boulocher C (2015) Intra-articular administration of xenogeneic neonatal Mesenchymal Stromal Cells early after meniscal injury down-regulates metalloproteinase gene expression in synovium and prevents cartilage degradation in a rabbit model of osteoarthritis. Osteoarthr Cartil 23 (1):122–133

Toh WS, Lai RC, Po Hui JH, Lim SK (2016) MSC exosome as a cell-free MSC therapy for cartilage regeneration: implications for osteoarthritis treatment. Semin Cell Dev Biol http://dx.doi.org/10.1016/j.semcdb.2016.11.008

Wang HS, Hung SC, Peng ST, Huang CC, Wei HM, Guo YJ, Fu YS, Lai MC, Chen CC (2004) Mesenchymal stem cells in the Wharton's jelly of the human umbilical cord. Stem Cells 22(7):1330–1337

Wang L, Tran I, Seshareddy K, Weiss ML, Detamore MS (2009) A comparison of human bone marrow-derived mesenchymal stem cells and human umbilical cord-derived mesenchymal stromal cells for cartilage tissue engineering. Tissue Eng A 15(8):2259–2266

Whyte GP, Gobbi A, Sadlik B (2016) Dry arthroscopic single-stage cartilage repair of the knee using a hyaluronic acid-based scaffold with activated bone marrow-derived mesenchymal stem cells. Arthrosc Tech 5(4):913–918

Advs Exp. Medicine, Biology - Neuroscience and Respiration (2017) 32: 123–124
DOI 10.1007/5584_2017

Index

CPSIA information can be obtained
at www.ICGtesting.com
Printed in the USA
LVOW05*1301271217
560944LV00004B/268/P